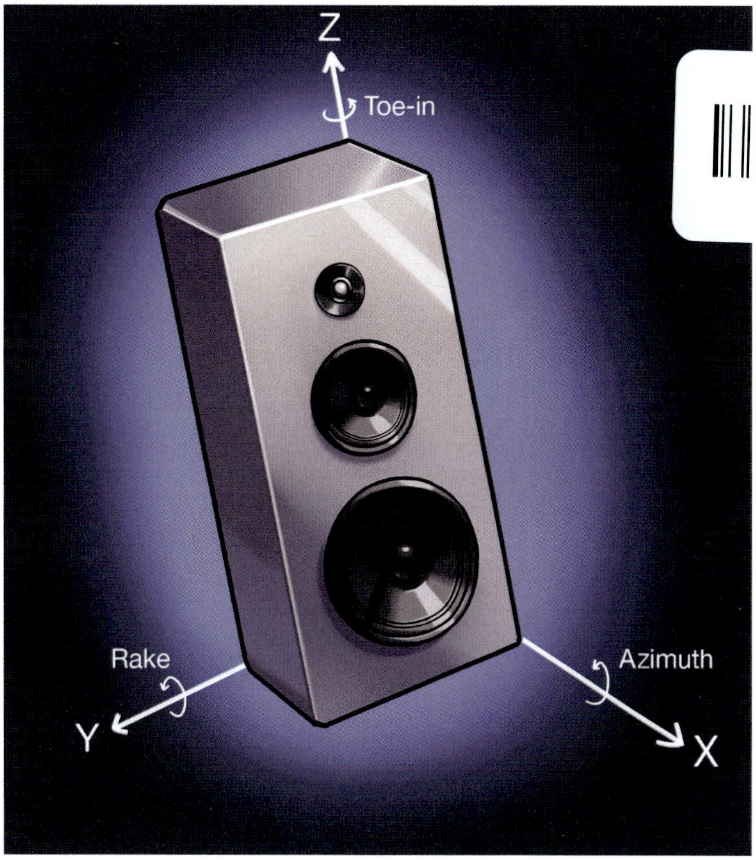

PRECISION SOUND

OPTIMIZE LOUDSPEAKER SETUP, ELIMINATE NOISE, AND DISCOVER MUSIC

BY TODD BINNIX AND STIRLING TRAYLE

ILLUSTRATIONS BY BRANDI WARREN

Inspired Forever Books
Dallas, Texas

Precision Sound: Optimize Loudspeaker Setup, Eliminate Noise, and Discover Music

Inspired Forever Books

Dallas, Texas

(888) 403-2727

https://inspiredforeverbooks.com

"Words With Lasting Impact"™

Paperback ISBN: 978-1-948903-99-8

Library of Congress Number: 2026902310

Printed in the United States of America

TABLE OF CONTENTS

FOREWORD BY STIRLING TRAYLE

It's essential to clarify the intent of this book. While we discuss many elements of system optimization, the core of this book is a detailed process for one approach to setting up a pair of loudspeakers to play music. It's incredibly challenging to articulate because this process is based on listening and lacks measurable criteria. A listening-focused loudspeaker setup requires discernment, and we have tried to equip you with various tools to make much of the process less subjective. Consequently, the recommended music tracks serve as tools to illustrate specific aspects of the speaker setup.

This book is not meant to serve as a definitive guide to loudspeaker setup. That would be absurd. It is part of an ongoing process that incorporates the experiences of countless individuals over the decades, some of whom I was fortunate to learn from, along with others who laid the groundwork for our understanding of audio reproduction. I have been afforded the opportunity to work on numerous music systems worldwide, and now it's my turn to contribute to the community of knowledge and share what I have learned and practiced.

We hope this book will spark discussion, debate, and improvement within the audiophile/manufacturer/dealer community. We would be delighted to find that others have taken the ball and run with it — improving upon our work and advancing the field as a whole.

After countless clients, rooms, and systems, setting up a pair of speakers has evolved into more of a flow of events rather than a step-by-step, goal-oriented, checked-off process. Because my clients hire me to work on their audio systems, I must respect their time and investment, so efficiency is paramount. Although the process isn't quick, I employ practices that avoid wasting time. But in my experience, no speaker setup proceeds swiftly.

A bit about this seemingly odd setup process: Many years ago, I realized that focusing on one speaker at a time would be easier for me than trying to keep track of two. This process uses a system that focuses on the left channel speaker first, and only when it is "done" does the right speaker begin to be moved "into place." The primary reason it is easier is that when two speakers are close to being aligned but not quite, their misalignment produces very distracting distortions that are difficult to pinpoint. I think of this as driver conflict.

While experimenting with this one-speaker-only concept, I tried turning off one channel and using my usual test tracks and techniques, but achieving meaningful results proved very challenging. Lowering the volume of one speaker helped slightly but didn't eliminate the problem—it was just a less intense version of the same driver conflict issue. The useful technique emerged when I fully misdirected one speaker by radically turning it and pushing it away from the listening position while keeping its volume the same. This positioning created an extremely diffuse and non-directional version of that channel, making it far less influential on the other. Granted, it still provided enough information to serve as a helpful anchor to which the other speaker could attach. *Surprisingly, stereo imaging becomes remarkably well-defined when a properly optimized left speaker plays alongside a deliberately misdirected right speaker.*

Around this same time, I started thinking of the speaker as a singular object in space; each axis of the speaker's movement presenting an opportunity to optimize its behavior in the room relative to the listening chair first, and to the other speaker second. The most surprising insight for me, while exploring how moving the speaker along each of those axes, was how even the smallest position or attitude changes could greatly impact the speaker's performance. I've since come to understand why this is and how to use it as a tool, but to this day, I still sometimes shake my head at how sensitive our ears and brain are when it comes to processing tiny differences in sonic information.

This setup methodology enables me to perform my job to the best of my ability. For me, it is my most powerful tool. I believe it will help you achieve excellent results with speaker setup as well. It requires a developed skill set executed to a high standard, and technical familiarity, with your speaker's behavior. I was surprised when Todd asked me to teach him this process. When he mentioned wanting to learn this for when I retired or ventured into the unknown, it made sense.

Todd, being the brilliant and curious individual he is, believes that investing in himself is a valuable way to navigate life. Going through the teaching process with Todd was not easy. It prompted me to think procedurally about how my workflow operates. Yes, I perform it the same way every time but breaking it down into specific steps and the numerous considerations that go into each step is (and was!) very challenging. Things we know intuitively often are hard to articulate and quantify. I stumbled through it with him.

In the end, Todd found his path to work through the process. But he did not take what he had learned and consider the journey complete. He spent countless hours practicing, making astute observations, and developing his listening ability, just as I have inadvertently done over the past forty years working in the audio industry. He will be the first to tell you that it's an ongoing learning process. Todd is a physicist, so saying he has an analytical mind would be accurate. He is also very matter-of-fact and practical. When he presented me with what could have been considered a well-thought-out and researched speaker setup book, I was taken aback. Here was a book about my setup process that I had no part in writing but had almost everything to do with the contents.

When I read it, I realized it was a book I could never have written. I greatly admire how he analyzed what I tried to teach him and transformed it into a thoughtful and well-researched book. Reading that first draft felt like reading a mostly filled-out crossword puzzle. In examining his evaluation and technical analysis of some aspects of the setup and the music material we use, I found myself saying, yep, that's what we're doing.

In some cases, I thought to myself, "Oh, what do you know? So that's what I was doing." It's been a surreal experience. Most of what you are reading consists of Todd's words because this is largely about Todd's journey. Had I been involved from the beginning, this book would not have been as clear or helpful as it is now. What Todd and I are trying to convey is quite complicated and requires the best of your passionate and dispassionate self.

As I've already mentioned, what is written here builds upon the efforts of countless other individuals and companies, whose research, development, products, processes, and simple curiosities have shaped our personal view of system optimization, with particular emphasis on loudspeaker setup and optimization. I hope Todd's journey and my experience, notes, and commentary provide a new way to approach how you can best enjoy music at home.

AN APPEAL TO SPEAKER MANUFACTURERS

Let's begin by recognizing the talented individuals who design and build the loudspeakers we enjoy. Creating a great speaker is no small feat—it requires engineering every component (cabinet, drivers, crossover, wiring, connectors, and feet) to work as a cohesive system and then bringing that design into consistent production.

Yet much of this effort goes unrecognized outside the manufacturer's walls. Consumers often think, *"That's not my problem,"* while manufacturers may focus so narrowly on design perfection that real-world listening environments are overlooked. Some designs, in fact, seem to assume the speaker will operate in a vacuum. Though design choices rightly belong to the manufacturer, added consideration for in-room behavior would be greatly valued.

These points may invite debate, but the repeatable process outlined in this guide deserves attention from even the most thoughtful skeptics. Our goal is to bridge the gap between design intent and actual performance.

Ultimately, it is the consumer—not the manufacturer—who must realize a loudspeaker's full potential. Performance depends less on a design's ideal conditions and more on how the speaker is set up. This guide offers a clear path to help users maximize their investment.

That reality, however, places an important responsibility back in the hands of manufacturers. If end performance is determined in the home rather than the lab, then certain design choices can either empower—or hinder—the user's ability to achieve it. While no loudspeaker can be made "setup-proof," thoughtful attention to how a product interfaces with real rooms and real floors can dramatically improve the likelihood that its full capability will be realized.

Two features are especially critical: adjustable feet and a reliable reference plane.

Starting with the feet: in many designs, they feel like an afterthought. Too often the threaded portion is overly short—sometimes only 3/8"—making it difficult to level the speaker precisely or achieve proper height and time alignment.

In our experience, at least **5/8" of threaded material** is necessary—more is even better. This provides enough range for fine adjustments while keeping threads safely engaged, which is especially critical in heavy speakers where stability matters. Equally important is **thread pitch**. Coarse standards like M8x1.25, M10x1.5, 3/8"-16, or 1/2"-13 may work for fastening but are poorly suited to precise leveling. A finer pitch allows more accurate positioning and can help push top-tier speakers to their peak performance.

The second key feature is a **reference plane**. A flat, consistent surface is essential for leveling a speaker in both directions. A flat cabinet top is ideal, but many modern designs use rounded tops and sides—whether for aesthetics or diffraction control—leaving only the bottom as a potentially flat surface. This makes consistent, repeatable alignment difficult. A clearly defined, flat reference surface tied directly to the driver array would be a game-changer for precision setup.

Manufacturers don't need major redesigns to achieve this—just the recognition that a simple, measurable reference point adds significant value. Such attention to detail could improve customer satisfaction and, in turn, sales, as users would find it easier to unlock the full potential of their speakers.

For those building traditional rectangular cabinets: good news—your designs already lend themselves naturally to precise setup.

To be clear, this is not an attempt to dictate how manufacturers should design loudspeakers. Whether to include adjustable feet or a reference plane remains their choice. Our intent is simply to highlight features that would help customers realize even greater enjoyment from these carefully engineered masterpieces.

PREFACE

Does the above picture resemble you? After spending a significant amount of money on his stereo system, he sits there trying to understand why he feels so dissatisfied with it. It certainly looked like me several years ago.

We all love music and are desperately trying to recreate an experience in our homes—connecting with the musicians. We're trying to understand why they wrote that specific piece of music and experience the emotion they're trying to elicit. My system always fell short of establishing an emotional connection with the music or the musicians behind it.

Like most audiophiles, while exploring the musical possibilities with a two-channel stereo system, I owned several different versions of each component—speakers, amplifiers, pre-amplifiers, CD players, DACs, and more. Each time I made a change, I hoped to achieve a better, more engaging music experience. Unfortunately, many of these decisions were based on some magazine or online review filled with flowery descriptions: improved soundstage width, enhanced depth, warm but detailed vocals, ruthlessly revealing, and so on. I thought if I could find the right combination of components, the music would pour out and magic would happen—it never did!

The process of changing equipment to find musical fulfillment is like assembling a 10,000-piece puzzle with no reference picture, grabbing pieces randomly in an attempt to create a coherent image. One might eventually piece the puzzle together despite many frustrations and wrong turns, but most would likely give up. In my case, musical engagement was elusive not because of the expensive, well-curated equipment I owned. I didn't need another trip to the stereo store. What I needed was the skill to extract the best possible experience from the equipment I already owned.

Consider this: if we're on a journey, how do we move toward our destination without knowing where we currently are? We need to understand where our system currently stands if we're going to achieve our goal of musical fulfillment. Optimizing an existing system will reveal the maximized performance of the equipment we already own. Then we know where we are. An optimized system is a form of ground truth. It's a foundation to stand upon so we can decide which changes, if any, we wish to make. We could be very content with our optimized system and decide to stay—or we could choose to continue the journey. Moreover, once our system is optimized, we will be able to hear clearly what any new component is doing and what is still missing.

As you will see in the chapters of this book, optimizing a system consists of carefully building a collection of "ground truths" so that we know where we are with ever-increasing accuracy and confidence. For example, we position a single speaker first and only address the second speaker when we've learned what we need to know from the first. Similarly, rather than working with musical tracks chosen at random (or based on our personal taste, which is effectively the same thing), we work with specific tracks that have specific things to teach us about the state of our system. In this methodical way, we work through the variables until we have a system that we truly know sounds as good as it possibly can.

This search for answers naturally prompts a deeper question: *What does our brain require to experience music as coherent and emotionally convincing?* When we understand that, we can—with great precision—identify the missing piece(s) needed to complete the sonic picture.

To achieve a high level of engagement, the brain must be able to relax and immerse itself in the musical performance. We want to believe that real musicians are playing in our room, but when the system is poorly optimized, it gives our brains many signals that something is not right. This results in a form of cognitive dissonance. When dissonance is present, the brain shifts into analytical mode, working to process inconsistencies, and rationalize rather than simply receiving the music. We begin thinking too hard, trying to imagine musicians in the room instead of effortlessly perceiving them.

Removing this cognitive dissonance enables a deeper emotional connection to the music. By optimizing stereo system setup, we create a low-noise, coherent sound field that dissolves those perceptual barriers and allows the listener to engage fully and naturally with the performance.

Complete and perfect dissolution of perceptual barriers is, of course, not possible in the real world. For an audiophile, realistic goals might include: (i) getting the best possible experience from the equipment they already own, (ii) making wise and informed decisions about any new equipment they choose to buy, and (iii) feeling secure about their system's performance so they are able to confidently judge the performance quality of various music tracks.

Unfortunately, optimizing stereo systems seems to be a dying art. Once upon a time, long before the internet and the commercialization of the CD, everyone had a turntable that required hands-on attention if an audiophile wanted to get the best from it. The need to pay careful attention to the turntable helped foster a willingness to pay careful attention to the entire system. Sadly, as the years passed, and turntables became less common, the "plug and play" mentality began to take hold.

A stereo system engineered for true high-fidelity performance is anything but plug-and-play. It takes skill and effort to optimize. Where can we find this knowledge and skill? I searched the internet for something that would make my system sound like music. Sadly, while there's abundant information online, not much of it is particularly useful.

Most websites or videos discuss the subject too vaguely or prescribe mathematical formulas that likely won't work in your specific room. Setup CDs or articles describe what you should hear in their reference songs but don't tell you how to position speakers to achieve that sound. What about a dedicated room full of acoustic treatments? This is more of the plug-and-play mentality—the idea is that the room is the problem, and if we hire an acoustician to prescribe absorbers and diffusers, all will be right. While that may improve some aspects of the sound, it further highlights the system's lack of engaging music.

To push a system to peak performance, we need to know exactly what we're trying to optimize. What does this mean? How do we know when we get there? And how do we optimize along the way?

This guide seeks to provide clear, actionable instructions. It's written for those who want to get the most out of their stereo system. If you're reading this, you're likely seeking ways to elevate your musical experience. You probably believe your speaker setup could be improved. And you'd be right to believe that. Optimizing your system and, in particular, focusing on speaker setup truly is worth the effort.

I have personally experienced everything described in this book. I aim to convey what, in my experience, works to improve a stereo system and enhance musical performance. I will describe many aspects of what I hear. This will be somewhat subjective, but I've done my best to objectively describe my experiences so you can identify the result when you hear it. Stirling and I seek to be as precise as possible in these descriptions so you, the reader, can grasp the concepts.

You may find some statements contentious. That's okay. If there's something you haven't experienced yet or don't believe in, feel free to ignore that part. The instructions on speaker setup are the core of this guide. Following them will enhance your listening skills and provide a more engaging musical experience.

You will notice snippets from sheet music to refer to when discussing certain songs. Reading music isn't required to understand the speaker setup instructions, but I thought including the sheet music might provide helpful insight for those who can read it.

Going through the speaker setup section requires time and patience. This isn't something you can master in just a few hours. Setting up a pair of speakers typically requires at least an entire day, even for someone proficient with this setup method. Before moving the speakers, you need to learn what to listen for and identify key elements in the reference songs. I have listened to all the songs thousands of times, and you will need to become very familiar with them as well.

The content of this book is very dense and rich with information. I don't consider it "casual reading." It will introduce ideas that are likely unfamiliar. This being the case, you will need to stop, think, and re-read some sections. That is normal. The good news is that you can take system optimization and speaker setup in bite-sized chunks.

The speaker adjustments detailed in chapters 8 and 9 are described in a step-by-step process. Although, you can pick any one of them, apply the instruction to the speaker(s), and the sound will improve. In this way, you can explore and learn each setting piece by piece.

For those who want to jump in and start adjusting speakers, a good place to start the optimization journey is with the right speaker. The full process always starts with the left speaker. But, while you read and digest the rest of the book, you can work with your right speaker to gain immediate rewards. You need to learn to listen effectively, become acquainted with the setup songs, familiarize yourself with your room, and practice the setup process anyway. So, working with the right speaker will get you started. To begin, skip to chapter 7, titled Right Speaker Exercises.

Keep in mind that the left speaker **establishes** the sound's foundation and since your left speaker is almost certainly not in its optimum position, going through the process of aligning the right speaker will make some things worse. But **matching** the right speaker to the left will also improve many things. The point is that you start to learn what to listen for and adjust the speaker's spatial and temporal presentation. Once you have experimented, you can revisit the process and repeat it with the left speaker, followed by the right speaker. Then you can optimize the noise floor and restart the process from the beginning. Each time you go through the process, you learn more about your speakers and your room. This will result in increasingly better sound and more musical engagement.

INTRODUCTION

THE SEARCH FOR MUSICAL TRUTH

The Philosopher and the Physicist — The Perspective of One Who Seeks

I want to take you back to the midpoint of my audio journey. This inflection point marks what I would consider the *real* beginning. Before this time, I did what most audiophiles seem to do—spend increasingly greater amounts of money changing components, with no real direction, in hopes of finding the magic. The shift in my thinking started when I changed jobs and encountered another audiophile at work. Of course, he invited me to his house to hear his system. After some chit-chat, he led me into his pitch-black, dedicated listening room and sat me down in the center chair. I could not see anything. He played several songs for me, and I was astonished at how amazingly good everything sounded. Vocals were incredible. It was as if the singer were right there in the room with me. When he turned on the light, I was stunned. I wasn't listening to a mega-dollar system. It was very humble, costing a fraction of what I had spent, yet offering many times the performance. What Paul did to achieve this was pay attention to detail. He worked on his system, trying to get every last drop of performance. Unfortunately, I was convinced that the room itself must have been the missing ingredient in my system, so I set out to recreate that experience in my own home.

The following year, I consulted an acoustician to design a dedicated audio room. I thought this was precisely what I needed and the final step in completing the audiophile journey. Was I ever wrong! I built the room, put in the acoustic treatment, and set up my system. Imagine my frustration when I could not get great sound and engaging music from my expensive audio equipment, which was sitting in a custom-built, acoustically treated room. I thought I was doing exactly what I was supposed to do, and what science says to do.

Conventional wisdom says we should purchase great components that work well together, build a dedicated room, and place acoustic treatment as an acoustician recommends. Conventional wisdom also says a bit about speaker setup. I knew most of this, like room ratio formulas. I knew the speakers should be symmetric and at the same distance from the center. I knew speakers should be pulled out from the wall. I knew they needed toe-in. I knew there should be a centered image and how to get it. All of this was insufficient.

Beyond this broad knowledge, I was at a loss on how to get the speakers to play engaging music. I desperately searched the internet, looking for anything related to speaker setup. I read books on room acoustics and took thousands of acoustic measurements. I continued to purchase more acoustic treatments, as that seemed to be the recommendation from books and acousticians for resolving sound issues. The more I added, the less satisfied I became. The question raging through my brain was, "What now?" Reaching that dead end made it clear that I needed a dramatically different approach—and someone with the experience to reveal it. With no clear path forward, I turned to the one option I had never tried: letting an expert show me what was actually possible.

Out of frustration, I hired Stirling to come and set up my speakers. Honestly, I wasn't sure what to expect. I read a review of his work, but it is difficult to know, based on a few pages in a magazine, just how transformational this process really is. After speaking with him, I was ready to see if there was one last card to play. Little did I know how working with Stirling would transform my understanding of what is possible with two-channel stereo.

Stirling showed up promptly, and I was with him for the duration of both 12-hour days. I tried not to bother him too much and mainly just observed. The first day was all about getting a lower noise floor in the system. I was initially skeptical about this part of the process, as it seemed he was simply cleaning the connections and leveling the rack and equipment. I was thinking, "Okay, it's great that everything is nice and clean, but how is this going to help?" At that time, I was almost entirely oblivious to the sound of noise and the damage it does to music. This is a topic that receives far too little discussion in the audio press. When he finished, I popped in a CD I had been listening to frequently at the time. I was surprised by how much more precise and clean everything sounded at the end of that first day. What stands out in my memory is how much crispness the guitar strings now possessed. It made a lasting impression on my mind, as I can recall it years later as if it were yesterday.

On day two, Stirling worked on positioning my speakers. He sat down and started to listen. Then he suddenly stood up and looked around the room at all of the acoustic treatments I had added. He said, "We need to remove as much of this as possible." After some discussion, I agreed, and we took all of it out to the kitchen and stacked it there. There were a lot of tube traps, panels, and various other items. I thought, "What will I tell my wife?" (If you're an audiophile, you've probably had the exact same thought at some point.)

Once the room had been returned to its basic configuration, he listened again and was ready to proceed. He had a very unusual method that I had not seen before. He pushed the right speaker back and turned it at about a 45-degree angle. He worked on the left speaker for most of the day. From about 9 a.m. to 6 p.m., he would intently listen to a part of one of several reference tracks, then go over and make very small adjustments to the speaker. After about the second hour of his work, I could no longer see the speaker move when he made adjustments (because the movements were so small). Occasionally, I would ask a question, or he would ask me to listen and make some comments. I recall learning just a few things in that first session.

1. With Nils Frahm's *Chant* playing, we hunted for the right distance from the front wall, and I soon realized how few positions actually worked. The goal was to hear the "growl"—a sign we'd found a good candidate—though I didn't yet grasp what that word really meant.

2. Establishing image height using *Mining for Gold* by the Cowboy Junkies. The singer should sound centered and be of appropriate height for a person standing between the speakers. Also, this track has no sibilance.

3. How much soundstage was possible and how organized the sound was with only the left speaker pointed at the listening position. I was astonished by how good and complete this sounded. How was this possible?

4. Making very small azimuth adjustments at the very end to fine-tune the sound. I was amazed at how much he could fine-tune a vocal track by making such tiny movements to the threads of just the right speaker.

I still remember listening to the first few tracks when Stirling was finished. I had never experienced stereo sound like this. The sound was so coherent, so dynamic, and so engaging. The sound of a piano filled the room, as if a real piano were sitting between the speakers. Symphonic pieces sounded as if the orchestra were right in front of me. It was really beyond what I thought could be done.

Many audiophiles focus on differences in sound. They seem to enjoy driving the system in different directions to experience what lies at the end of each path. What Stirling did to my system was not just different; it was better in every regard. It was profound and correct, and it was music.

NOTE: Some of you may be familiar with the "Sumiko" or "Masters Set" method of speaker placement. Stirling Trayle and John Hunter were the originators of this speaker placement method, which could reasonably be considered a very, very basic method of speaker placement. It was developed to help audio dealers quickly place speakers for acceptable results. What's presented in this book may appear to be an evolved, highly advanced version of that approach. In truth, however, it only resembles the "Sumiko" method in one respect: we begin by working with the left speaker first.

I listened to this system untouched for well over a year. Then a dealer called and said he had an excellent offer for me. I could upgrade to the next speaker model up for a modest amount. I traded my existing speakers and bought the "upgrade." This was a huge mistake. The new speaker never really worked in my room. I missed my old sound. No matter how hard I tried to achieve the same type of sound that Stirling had, I lacked the necessary skill to achieve good results.

At this point, I contacted Stirling and asked if he would be willing to come out and spend several days teaching me how to optimize speakers while also determining where the new speakers would sound their best. He agreed.

That first visit was mostly about listening—how and what to listen for. We reviewed many of the reference tracks and discussed various aspects of the sound. We started similarly to the last visit with the right speaker turned away. I honed my skills using Nils Frahm's *Chant* by focusing on the sound of the growl and the piano's right hand. Many enigmatic things were happening that I couldn't comprehend at the time. I will provide a few examples that seem funny now but were a challenge for me to grasp at the time—not only why they were important but also what to do about them.

These early moments set the stage for the first lessons that finally began revealing what I had been missing. We were using *Chant* and finding some speaker positions that seemed initially positive. Stirling would say, "We need to move on." But why? I would ask what he was doing, and he would respond, "Do you hear there is no sound coming from the other side of the room?" I would say, "Oh, I hadn't noticed that." Then he would say something like, "Well, that's not good. We need to fix that."

Another example I distinctly remember was when we found a good speaker position where *Chant* worked. I returned to the listening chair and pressed play on Shostakovich's First Symphony. Literally, within three seconds, he said, "Stop! Listen to that." I said, "Seriously? Listen to what? The music hasn't started yet." He said, "This is the first time the hall has come up." What? That is when I learned about that aspect of sound and used the hall's sound as a tool to understand where the sound is and where the sound isn't.

Another thing I recall learning in that first session was the concept and application of sound pressure on the body. This one is funny. Stirling said, "I want you to feel the sound with your face. Feel the pressure on your face and notice where it is and where it isn't." This sounds simple. However, we were listening to the opening of Shostakovich's First Symphony. This is not some bass-heavy, chest-pounding track. The pressure is very light. I was thinking, yes, I can feel pressure… I think. It is here and here.

Then he moved the speakers some, and he said, "Where is the pressure now?" I intently tried to feel the pressure on my face but didn't notice anything. I thought I wasn't getting it because there must be pressure, but I can't feel it. So, in frustration, I said, "I don't feel anything." He said, "Right! There is no pressure here." Damn it. He tricked me. But really, he was trying to teach me that if you find a spot and you don't feel this pressure, then it's not a good position. At the time, I was just learning about the idea of pressure. I had no idea how to apply it at this point. Amid these experiments, something unexpected happened—an exchange that showed me just how unusual these perceptions can feel.

At some point during this lesson, Stirling turned to me with a very sincere look and said, "You have no idea how relieved I am. I wasn't sure if anyone else could feel this or if I was just crazy." I am sure I learned other things in that first session, but by the end, I think I was more confused about how this setup system worked. I had learned stuff, but I had no idea how to apply it. Stirling left, and we scheduled another visit in three months.

In the meantime, I tried to practice some of what I had learned. I listened to the reference songs every day. I tried to let my mind process all the information I had gathered and organize it into something usable. It was rough. I wasn't any better at setting up a pair of speakers than I had been before. It is one thing to know what to listen for and be able to identify defects in the sound field. It is another thing to utilize that information and know what adjustments to make to the speakers that will correct those defects.

The ongoing struggle made it clear that I needed more guidance, which led directly to the next round of training. Finally, the day arrived for the second set of lessons. During this visit, we worked more on listening skills. On the first day, we worked with *Poem of Chinese Drum*. Stirling was quite busy working on the speaker, and I asked, "What are you doing?" He said, "I am trying to make the drum round." I thought, "What? What is this guy talking about? The drum is round." I sat and listened, and he said, "I am not talking about some imaging thing."

Okay, if not imaging, then what? At some point, we were sitting there, and Stirling asked me to listen and visualize the sound in my mind's eye. I did that, and this is where we discovered that I could hear in a different way than Stirling.

When listening to the left speaker, I could toggle my focus from the broad sound in front of me to listening specifically to the left speaker. When I focused on the left speaker, I could "see" the sound in my mind. As various words were sung or notes were played, I could see what they were doing. I could visualize them as though they existed in a circle around the speaker. I could see how a specific note would shoot up, or to the left, or diagonally, or forward, etc. These words and notes moved around inside the circle as the song played. I compared it to watching the old arcade game *Tempest*.

At that point, I knew what Stirling meant by making the drum round. At least, in my mind, I knew what that meant. It turned out that Stirling heard it differently, but he meant the same thing, and the same thing needed to be fixed with the speaker placement. The idea was that if we heard part of the sound jetting out in a certain direction, then we needed to move the speaker in that direction to "capture the sound." We wanted to hear all the sounds in a nice, tight circle with nothing shooting out in any direction. This is the definition of coherence.

The concept that sound could be "well-behaved" was probably the most important insight I gained during the second session. It gave me a systematic way to determine which of the six directions a speaker needed to move based on what I was hearing. During that session, we began focusing more closely on identifying timing defects in music. We spent extended time listening to several tracks—particularly "Like the Weather," "Sleeping Lessons," and "Heart of Ice"—analyzing their timing relationships in detail.

I vividly remember what I learned from Joe Jackson's *"Heart of Ice."* The track features a drummer rhythmically tapping a hi-hat throughout. About twenty seconds in, a flute enters. Stirling pointed out that the timing between the hi-hat and the flute was inconsistent. "Do you hear how the hi-hat shifts from emphasizing the downbeat to the upbeat?" he asked. I told him he was imagining things. After all, I reasoned, the same drummer played the same part—how could the timing change? It was recorded onto the CD! How could playback alter it?

We replayed the section several times. Eventually, Stirling suggested I tap my finger in time with the hi-hat as we listened. To my surprise, once the flute entered, I couldn't keep the beat. The rhythm stumbled—it wasn't my imagination. Once I heard it, I couldn't unhear it. The experience revealed that speaker setup could distort musical timing, and that precise adjustments could correct it. With those puzzles finally beginning to make sense, the broader lessons of the session came into focus.

Coming out of that session, I had a much clearer understanding of several aspects of speaker setup, though many ideas were still swirling in my mind. I wasn't yet confident enough to perform a complete setup on my own, so I practiced with an old pair of speakers, refining what I had learned. Still, it felt like only part of the puzzle. I decided to take the next step. I bought a well-regarded pair of used speakers and a top-tier preamplifier, setting them up in my untreated front room. Then I arranged a third session with Stirling—about six months later—to explore how system changes and room acoustics affected performance.

That third session tied many threads together. I gained a solid grasp of finding optimal toe-in and rake angles and using *"Chant"* to locate the best front-to-back position relative to the wall. I also deepened my understanding of lateral positioning. A new focus was on using the piano in Beethoven's *First Piano Concerto* to fine-tune the left speaker's forward, lateral, and rake adjustments to "window out" the room. The work involved subtle movements—tiny bumps or minute turns of the spikes—and, for the first time, I was the one making the adjustments.

This session also reinforced the importance of *pressure* as a listening tool. We paid close attention to where on my body and face I could feel sound pressure and what those sensations revealed about the speaker's position. By the end, I had developed a functional understanding of how to set up a pair of speakers correctly. While questions remained, the fundamentals had finally clicked into place.

We also explored how changing components or cables affected system behavior. For instance, we began with a good preamplifier. The sound was well organized and musically satisfying. Then we replaced it with a higher-end model—one that cost five times more. The result was startling. The piano became more articulate and transparent, a clear sign of the new preamp's lower noise floor. Yet the orchestra suddenly sounded disorganized, almost chaotic—more like a student ensemble than a professional one. Timing had collapsed.

To correct this, we went back to work—making a series of small lateral and forward bumps, followed by compensating toe-in and rake adjustments. Gradually, the sound was reassembled. When the music locked back into coherence, the improvement was remarkable: greater clarity, precision, and articulation than before. The lesson was unmistakable. You can't simply insert a new component into a system and expect optimal results. Every change requires realignment of the speakers to accommodate the new component's behavior within the system's ecosystem.

This realization overturned years of "audiophile training." I had always believed speakers were set up once and left in place while components were swapped in and out for evaluation. That assumption, I now knew, was completely wrong. This realization reshaped how I approached every system change going forward.

After that, I finally had enough understanding to begin learning independently. The puzzle pieces were in place, and I could start seeing the pattern for myself. Stirling still answered my questions, though often with cryptic, almost Zen-like responses—"Have you considered the sound of one hand clapping?" Okay, maybe not quite that obscure, but always designed to make me think. With each system I worked on, my insight grew. After tuning a couple of dozen systems, I could identify new tracks that exposed specific setup issues just as clearly as the originals we had used.

I learned to recognize noise and its influence, to perceive what individual components contributed—or detracted—from the overall sound. Introducing a new component was no longer a mystery; I could hear exactly how it changed the system, for better or worse.

So why tell you this story? Because you can experience this transformation too. Once you hear a truly optimized stereo system, it's intoxicating—almost addictive. After my first exposure, there was no going back. I realized anyone can swap components, but few people can truly *maximize* a system. Stirling was one of the rare few who could. But even he won't be doing this forever. That thought pushed me to learn how to do it myself and to share it with you.

Honestly, I didn't expect the process to be this challenging. I thought a few days of instruction would be enough. Instead, it became a multi-year journey that demanded persistence, patience, and humility. System optimization is not something you master in a weekend—but if you persevere, the reward is immense. Eventually, you'll be able to diagnose and fine-tune any system, anywhere. When you visit an audio show or demo room, you'll listen with an analytical ear and know precisely what's right—and what isn't.

Sometimes I wish I could go back and hear that first system Stirling set up for me—the "before" version, the one I thought sounded reasonably good at the time. I'm certain that, by my current standards, it would sound shockingly poor.

Summary: Where the Journey Begins

The experience with Stirling marked a turning point—not only in how I listened, but in how I understood the very concept of listening. What began as a search for better sound gradually became something deeper: a study in perception, patience, and physics. Every movement of a speaker, every change in tone or texture, reflected an underlying principle of how we, as humans, process sound.

That realization leads us to the starting point of this book. Before we move speakers, evaluate recordings, or attempt to reveal the full potential of a stereo system, we will first understand the instrument through which all music is experienced: the human auditory system.

Section I lays the scientific and perceptual foundation for everything that follows. In the chapters ahead, we will examine how sound is processed by the ear and brain, how music itself is structured, and why seemingly small disturbances in timing and coherence can cause a stereo system to fall short of musical truth.

To move forward on this journey, we should first consider what our ears and brain are actually doing. Before we can master speaker placement or perceive true coherence, we will understand the mechanics of hearing such as the way vibrations become perception, and how the smallest timing differences translate into spatial and musical meaning.

The next chapter begins there: with the science and wonder of how we hear.

SECTION I

FOUNDATIONS OF MUSICAL TRUTH

Why This Section Matters

Before you can place a speaker in the right location, you should first understand **what you are hearing** and why the speaker behaves the way it does. Speaker setup is not random trial and error—it is the deliberate shaping of sound, timing, and energy inside a three-dimensional space. To do that well, it's important to know how your *ears*, your *brain*, and your *room* process sound.

Section I provides that foundation.

Modern systems reproduce music with extraordinary capability, yet most listeners have never been taught what actually happens as sound moves from the speaker to the room to their body and finally to their mind. They hear differences—brightness, muddiness, smeared attacks, center images drifting left or right—but do not know *why* these problems occur or how to correct them.

This section fills that gap.

You will learn how the human auditory system extracts meaning from timing cues only microseconds apart. You will see how noise—electrical, mechanical, and acoustic—quietly erases the very details that make music feel alive. You will discover why the "space between the notes" matters as much as the notes themselves, and how your brain reconstructs musical patterns even when information is missing or distorted.

Most importantly, you will begin to view your audio system as a **musical instrument**—one that requires calibration, precision, and care. Every chapter in this section builds the mental model you will rely on for the rest of the book. When we later begin moving speakers in tiny increments—sometimes as small as the thickness of a fingernail—you will know exactly *why* such movements matter, and *what* your ears should be listening for.

Think of Section I as the lens through which speaker placement becomes intelligible, predictable, and repeatable. Once these concepts click, the entire process becomes far easier—and infinitely more rewarding.

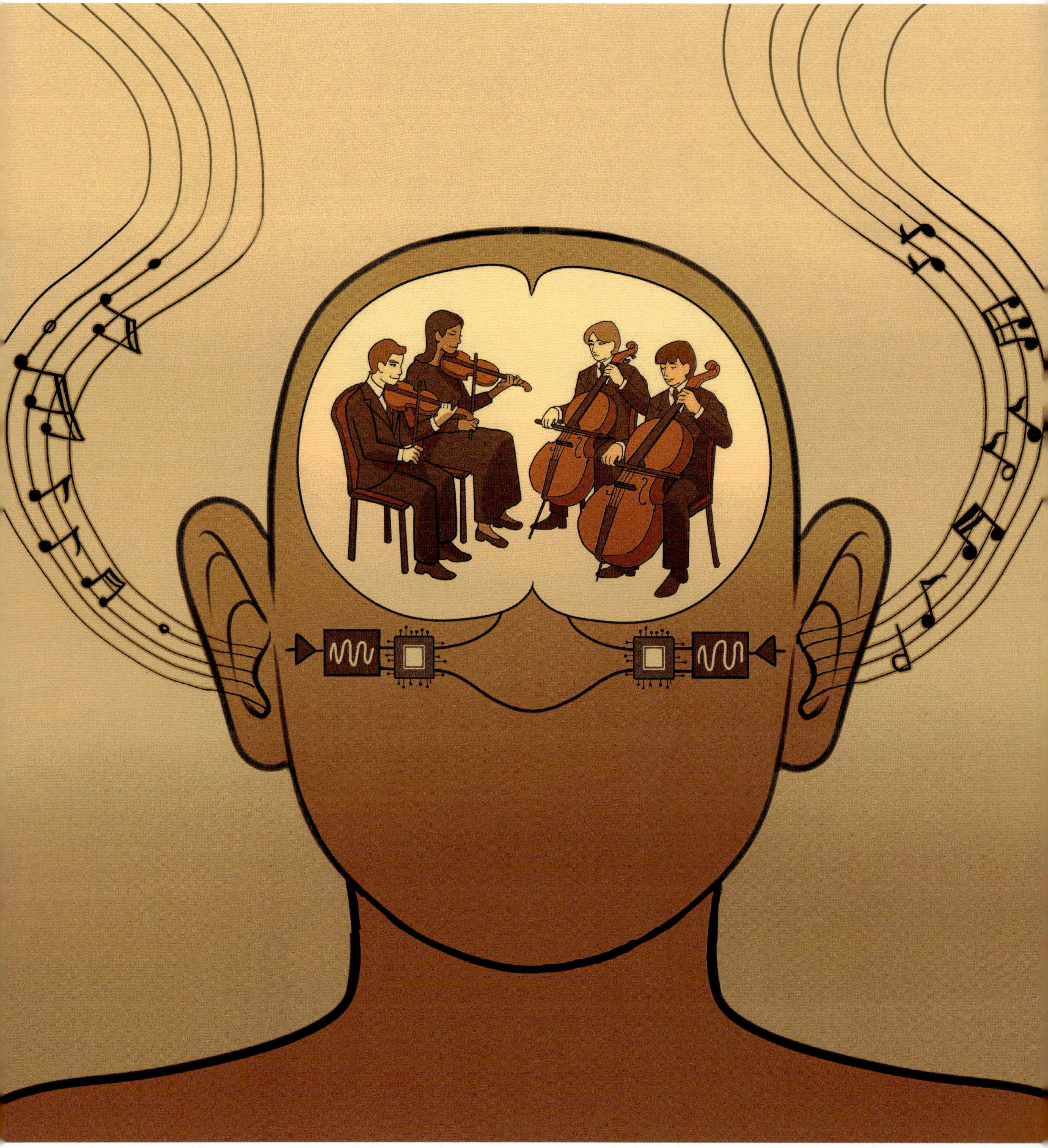

CHAPTER 1

PROCESSING SOUND — HOW WE HEAR

Human Hearing and Stereo Sound

Your ears and brain together form the most sophisticated audio processor ever created—capable of detecting timing differences under **10 microseconds** and spatial shifts less than one degree. Understanding how this system works offers valuable insight into the observations required for precise speaker placement.

Every serious book on stereo insists you *trust your ears*. But what, exactly, are your ears detecting when you move a speaker by an imperceptible amount? This chapter explores the science behind those perceptions, providing context for what you are hearing.

What is Sound?

Sound is a mechanical wave that requires a medium—**air, water, or solid matter**—to propagate. It moves through the vibration of molecules within the medium, and because these vibrations occur parallel to the wave's direction of travel, sound is classified as a longitudinal wave.

Like all waves, sound is defined by key properties: **speed, amplitude, frequency, phase, and superposition**. The human ear and brain draw on each of these to interpret and make sense of the world around us.

Specifically:

- Speed refers to the rate at which waves move through a medium.
- Amplitude determines loudness, reflecting the wave's energy. Higher amplitude waves produce louder sounds.

- Frequency corresponds to pitch, with higher frequencies perceived as having higher pitch.
- Phase refers to the relative starting point of a wave in time.
- Superposition defines how different waves interact with one another. This influences timbre (or tone), which is the quality of a sound that distinguishes different sources, even if they produce the same frequencies.

Each of these properties plays a fundamental role in how we perceive and experience sound. We will explore each in detail and relate it to the optimization of a stereo system.

One fundamental property of sound is its constant velocity in air, where it travels at 1130 feet per second (ft/s) or 343 meters per second (m/s). Converting this to a scale more relatable to our listening rooms gives 1.1 feet per millisecond (ft/ms). Sitting 10 feet from a loudspeaker, sound therefore reaches the ear approximately 9 milliseconds after the driver begins to move. While the speed of sound itself has no direct bearing on speaker setup—since it remains essentially the same in every room—the events unfolding during those 9 milliseconds are critically important. So what, exactly, is happening in that brief span of time?

When a loudspeaker driver moves outward, it compresses the air molecules directly in front of it, creating a pressure wave that propagates through the air toward the listener. This motion establishes alternating cycles of compression and rarefaction, in which air molecules bunch together (high pressure) and then spread apart (low pressure). These pressure fluctuations travel across the room and ultimately reach the ear, where they set the eardrum—a thin, flexible membrane—into motion. The resulting vibrations are transmitted through the ossicles (the tiny bones of the middle ear) and converted into electrical signals within the cochlea, which the brain then interprets as sound.

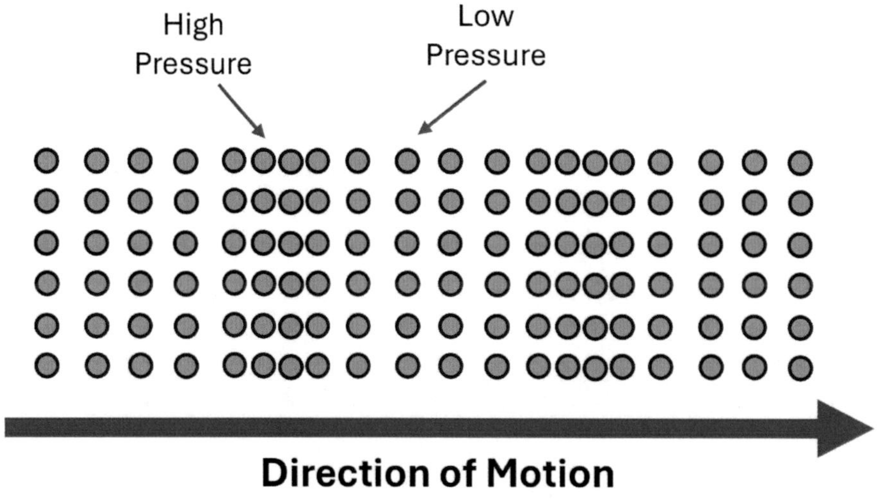

Each dot in the depiction represents one air molecule. The pressure (sound) wave moves from left to right. Notice where the lines of dots are closer together, as this represents the moment the driver compressed and formed a high-pressure packet. When the dots (air molecules) are closer together, the sound pressure is higher, and therefore, we perceive the sound as louder. If the air were visible, we would see many evenly spaced pressure packets traveling from the speaker toward us.

Music contains a wide range of frequencies at various amplitudes at any given instant. Imagine the organization of these air molecules into a highly complex pressure wave emanating from your speakers. Your speaker's drivers encode all of the various frequencies into a continuous pressure packet stream of information that our ears detect, and our brains interpret. Divide this stream into tiny increments of time so that the listener perceives a distinct pressure at a given instant. At that moment, the brain must integrate the wave properties into information regarding the sound's loudness, pitch, and direction. Imagine the complexity of this task that our brain performs naturally, again and again as the music flows.

In the late 1800s, scientists began seriously studying how the human brain integrates sound and interprets pressure waves. Several scientists have proposed various theories regarding the three aspects of our hearing: loudness, pitch, and temporal resolution. In his book, "Human and Machine Hearing: Extracting Meaning from Sound," Richard Lyon expertly summarizes the rich history of scientific research in human hearing over the past 150 years, encompassing a range of theories. Reading this book will provide you with a history of audiology and introduce modern theories of human hearing.

While there have been recent advances in theories of human hearing, we still don't know exactly how the ear, body, and brain work together to interpret sound waves and produce an "image" of what we sense. Yet, some interesting experiments have explored human hearing capabilities. These experiments provide insight for setting up a stereo system that reproduces realistic illusions of a music performance. We will examine pertinent experiments regarding tone, timing, and dynamics, beginning with the latter.

Loudness Resolution

The human ear is a highly sensitive pressure sensor. It can sense pressure changes from 20 micro-Pascals on the low end, which corresponds to a movement of the eardrum of approximately 1 picometer (1 trillionth of a meter). At the upper end where pain starts, the pressure is 20 Pascals. This is an extreme sensitivity range of over one million times. The extensive range encourages scientists to employ a common plotting trick that utilizes the logarithmic scale. This plotting device compresses the information into a more manageable range. It is known as the **Decibel Scale**, where a Decibel equals one-tenth of a Bel and is abbreviated as dB. It is also known as SPL, which stands for sound pressure level.

On this scale, zero decibels is considered the **threshold of hearing**, and 120 decibels is the pain threshold. A typical listening level for home music might be 80 dB. It is essential to note that prolonged exposure to high sound pressure levels can cause permanent hearing damage. Exercise caution when playing your stereo at high volumes (greater than 90 dB). For example, according to the National Institute for Occupational Safety and Health, spending more than 15 minutes at 100 dB can cause permanent hearing damage.

While the ear is highly sensitive to a wide range of pressure variations, its response is nonlinear. How loud we perceive a sound depends on its frequency. In 1933, H. Fletcher and W. Munson published "Loudness: Its Definition, Measurement, and Calculation." The study contains data curves showing perceived loudness versus frequency, which are now famously known as the Fletcher-Munson curves (shown here). The curves were later refined by D. Robinson and R. Dodson in a 1956 study. More recent studies have further refined the curves, resulting in the current ISO 226 standard.

Fletcher-Munson Curves

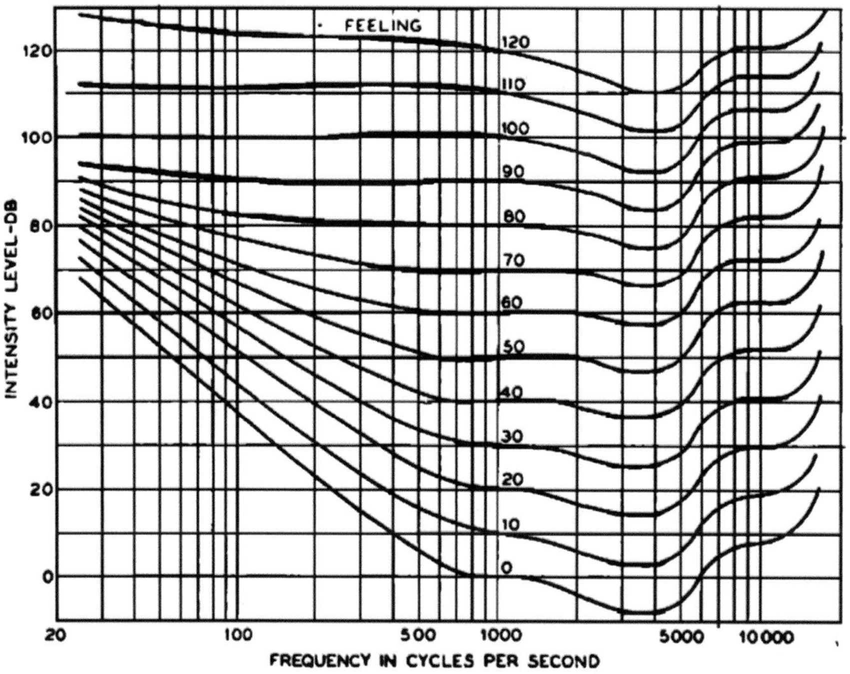

The curves shown in the graph above demonstrate that we become increasingly insensitive to sound pressure at the extremes of our hearing range. For example, as bass frequencies descend below 200-300 Hz, we require increasingly more pressure to perceive them as having the same loudness as a sound with the same pressure at 1000 Hz. Despite this non-linearity, we have adapted to understanding that this is how things around us sound. We will disregard this non-linearity when setting up a stereo system. We are not trying to compensate for it. We will position the speakers so that the sound they produce is normal to us. Although, two practical observations can be made from the family of curves.

First, the curves flatten out as the loudness increases, meaning our ears are more linear. Notice that the range from 80 to 90 dB is relatively flat up to 1000 Hz. This is a good range to listen for optimum resolution. The second observation is related. Some pre-amplifiers feature tone controls for listening at lower volumes. These can be used to boost the bass and treble, allowing for more linearity at lower listening levels.

Music composers utilize our extensive range of loudness resolution, marking the music score from fff (extremely loud) to ppp (extremely soft). The slow or abrupt changes in loudness imbue the music with anticipation and excitement. We are anxious about where we are going, but we look forward to getting there. Interestingly, music can be simultaneously beautiful and boring without dynamic contrast. For example, plainchant, which has no dynamic contrast, was used in the Middle Ages for prayer and meditation.

Noise intrudes on our ability to experience the full dynamic range of music. The everyday environmental sounds around us hamper the music's expressiveness. The refrigerator, the air conditioner, and cars passing by are all responsible for the extraneous sounds we endure while listening to our stereo system. The sum of all these sounds is called the **acoustic noise floor.** (This is in addition to the noise generated within the system itself.)

Loudness variation below the noise floor cannot be perceived. Thus, the noise floor robs the music of the composer/performer's intended feeling and excitement, replacing it with a homogenized sound. You can dramatically improve your music experience by reducing your noise floor. Ideally, in a normal home, the noise floor would be as quiet as a library, about 40 dB. A custom, or dedicated, room would typically be required to achieve noise floor levels below that level.

Frequency Resolution

Before diving deeper into how we perceive pitch, we need to clarify the terminology that underpins frequency perception. Examining how our brain interprets frequency information requires defining four key terms: frequency, tone, note, and pitch. The four words are related but have subtle differences.

- Frequency is the rate at which the sound wave oscillates. It is the number of cycles per second and is measured in units of Hertz. For example, the frequency of A4 is 440Hz. This is a scientific term.
- A tone is a sound recognized by a regular vibration. A simple tone has only one frequency, whereas a complex tone has multiple frequencies. The extra frequencies are called harmonics. The fundamental, which is the first harmonic, is the lowest frequency of a complex tone.
- A note is written or read. A tone is played or sung. Music existed long before Guido of Arezzo defined modern musical notation in his seminal work, "Micrologus," in 1028 AD.
- A pitch is the brain's perception of a tone's relative highness or lowness. A tone's pitch is similar to its frequency; however, audiologists distinguish between pitch and frequency. Therefore, they don't refer to a tone as having a pitch of 100Hz.

When considering frequency resolution, a key question to ask is: What range of frequencies can a typical human perceive? Many people have been taught that a young, healthy human's hearing range spans from 20 Hz to 20,000 Hz. These often-quoted figures originate from a 1922 paper by E.P. Fowler and R.L. Wegel, titled "Audiometric Methods and Their Applications," published in the *Transactions of the American Laryngological, Rhinological, and Otological Society.*

The study's experiments assessed the frequency range a person could perceive using simple test tones. Similar studies have been conducted over the years, generally yielding the same result. For example, a 1985 article in *Audiology* by E. De Seta, G. Bertoli, G.A., and R. Filipo, titled *High-frequency Audiometry Above 8 kHz: Comparative Results of Normative Thresholds Obtained with a Headphone System and a Quasi-Free-Field System*, confirmed the established limits.

In essence, humans can hear frequencies from about 20 Hz to 20,000 Hz, though individual differences exist, and the upper range tends to decline with age. Now, let's examine the two extremes of sound perception more closely.

At the **lower end** of the frequency spectrum, our experience of sound becomes dual in nature—we hear it and simultaneously feel it. Fowler and Wegel, in the paper referenced earlier, were among the first to document this phenomenon. Their findings show that this perceptual crossover begins at roughly 60 Hz. Below that threshold, we gradually transition from hearing to feeling the sound's physical energy. They noted,

"The data indicate this point (the lower limit) to be around 20 cycles per second. For frequencies lower than this, the sense of feeling is more acute than that of hearing."

The essential takeaway is that while we may not *audibly* hear frequencies below about 20 Hz, we absolutely *feel* and experience their presence through vibration and pressure.

Recent neurological studies have challenged and refined our traditional understanding of these limits. Modern research supports and extends these early observations. In a 2017 study presented by Christian Koch at the *12th ICBEN Congress on Noise as a Public Health Problem*, researchers used MRI imaging to explore how the brain responds to low-frequency sound. Sixteen subjects were exposed to pure tones ranging from 8 to 250 Hz using a custom, low-distortion, MRI-compatible sound source. The imaging revealed auditory cortex activation down to 8 Hz, and, importantly, no other brain regions were stimulated at that level of significance ($p = 0.001$; $N > 22$).

Koch writes:

> "Brain activity was found down to 8 Hz exclusively in the auditory cortex... The strength of the BOLD signal decreased with decreasing frequency starting at 250 Hz, reached a minimum around 20 Hz, and then increased again at still lower frequencies. These data and further measurements suggest that the perception mechanisms may change at around 20 Hz. In addition, they clearly show that infrasound induces a hearing process."

This means that at infrasonic frequencies—below the conventional threshold of hearing—the brain processes sound as **both** a *hearing* event and a tactile one. These findings help explain why the lowest frequencies play an outsized role in how music not only sounds but also feels.

The ability to feel bass energy is crucial to how we perceive and enjoy music. When analyzing the frequency content of orchestral recordings—or even small ensembles—we consistently find substantial energy below 20 Hz. Studies, such as those by Sean Olive in his 2004 paper, suggest that as much as 30% of our musical enjoyment derives from bass content. In the context of stereo reproduction, this low-frequency energy isn't just an accessory—it's a vital component of emotional engagement. Bass gives music its weight, presence, and physical connection to the listener.

The first step in the speaker placement method outlined in this guide is to identify the position in the room where we can feel bass energy below 35 Hz. Since our bodies are mostly water, we are particularly sensitive to this low-frequency energy, which resonates through us and is felt physically, often in the belly. Be aware, there is a distinct difference between hearing a fundamental tone at 35 Hz and feeling it. It's important to note that our brains can also infer the presence of the 35 Hz tone based on its harmonic content, which will be explained further in the next section.

On the other end of the **frequency spectrum**, at 20,000 Hz, the typical human hearing range diminishes with age. A young person can hear up to 20,000 Hz, but by middle age, this often drops to approximately 15,000 Hz. Again, it is essential to differentiate between hearing and *perceiving* frequencies. Traditional hearing tests in the 20th century simply asked subjects, "Can you hear anything?" as various tones were played through headphones or speakers. Today, we can use more advanced techniques, such as PET scans, to monitor brain activity and see how it responds to specific frequencies without relying on the subject's verbal response. One study that utilized this method, "Inaudible High-Frequency Sounds Affect Brain Activity: Hypersonic Effect," by T. Oohashi et al., published in the *Journal of Neurophysiology*, shows that subjects' brains respond to frequencies well above 20,000 Hz, even though they do not consciously hear them.

The **upper limit** of sound perception relates to stereo sound reproduction as it connects back to cognitive dissonance. In nature, sounds are produced with a rich spectrum of harmonics that extend well beyond 20,000 Hz. (For example, see the paper: "There's Life Above 20 kHz! A Survey of Musical Instrument Spectra to 102.4 kHz," by James Boyk.) It feels unnatural to have a "brick-wall" filter that cuts off all frequencies above an arbitrary limit—like the 22,050 Hz limit imposed by the Red Book standard for CDs. This is just a theory, but it helps explain why many audiophiles prefer the sound of SACD or high-resolution audio files. It may also shed light on why some people enjoy the effect of super tweeters, which extend the frequency response of their systems beyond 20,000 Hz. The unbounded high frequencies just feel more natural, and although they aren't the focus for speaker alignment, they are still important. Once your system is optimized, I encourage you to experiment with supertweeters. Who knows, you may find a key to unlocking another level of musical fulfillment.

With the practical boundaries of human hearing now established, we can turn from the question of *what* we detect to *how* we interpret it. This shift brings us naturally to an examination of tone and pitch perception. Scientifically, a tone is produced by a regular vibration and is defined by its fundamental frequency and its series of harmonics. A pure tone contains only a single frequency, with no overtones, whereas a complex tone contains multiple frequencies whose harmonics shape its timbre. Pitch, in turn, is our perception of a tone's fundamental frequency.

The definition of tone is rooted in acoustics and physics, where sound is analyzed regarding vibration patterns and how they interact with the human auditory system. Meanwhile, the definition of pitch is rooted in music. While you likely have an intuitive grasp of tone and pitch, it is essential to consider how the human brain processes tones and perceives pitch. This process is highly complex and has several fascinating aspects to explore.

From the mid-1800s through the mid-1900s, primarily due to the work of Ohm and Helmholtz, the ear's cochlea was viewed as purely a frequency analyzer—essentially a biological Fast Fourier Transform (FFT) machine. A key implication of this model is that the ear would be completely phase-blind. Granted, experimental observations contradict this idea. The ear and brain work together intricately, in a manner, far more complex than simply computing hundreds of FFTs per second.

In 2006, Probst proposed that the cochlea transmits neural impulse patterns that encode a sound event's nature and timing. This perspective aligns more closely with modern thinking on pitch perception. Notice how pitch and timing are interconnected—this makes sense because pitch is fundamentally a series of repeated pressure events over time. Whether it's a pure or complex tone, the timing of these events must be encoded.

How timing affects timbre is most clearly demonstrated by a famous experiment conducted by K. W. Berger in 1963, published as *"Some Factors in the Recognition of Timbre"* in the *Journal of the Acoustical Society of America.* In this study, Dr. Berger recorded various wind instruments—flute, oboe, saxophone, trumpet, clarinet, French horn, and others. He then modified the recordings using three types of filters, one of which removed the initial transient and trailing end of each note. Importantly, the frequency spectrum of the instruments remained unchanged. Trained musicians were asked to identify which wind instrument was playing. When the transient information was removed, the subjects experienced significant confusion and, in most cases, incorrectly identified the instruments.

In his book, *The Sense of Hearing*, Christopher Plack explains: "Frequency may be represented by the pattern of activity across neurons with different characteristic frequencies." In other words, the brain maintains a "map" of sounds, allowing it to compare new auditory inputs with existing patterns quickly.

Our brains rely on pattern recognition to process complex sensory information efficiently. This is particularly evident in visual perception, where we don't consciously analyze every detail of a familiar face each time we see it. Instead, our brain has constructed a mental "map" or template based on repeated exposure. This allows us to instantly recognize someone, even under different lighting conditions or viewing them from a different angle. We don't need to register every eyelash or wrinkle consciously; instead, we perceive a holistic pattern representing the person.

A similar process happens with auditory perception. Just as our brain forms a visual template for familiar faces, it also constructs auditory templates for familiar voices, sounds, and speech patterns. When we hear a voice we know well, we don't have to analyze each individual sound or syllable to recognize who is speaking. Instead, our brain quickly matches the incoming auditory pattern to a stored "map" of that voice, enabling near-instant recognition. This efficiency helps us navigate a world filled with complex sensory input while minimizing cognitive load.

Three Features of Pitch Perception

This ability to recognize patterns in sound—not just tone color but the structure of frequencies over time—also plays a key role in how we perceive **pitch**. With that foundation in place, we can now look at three important features of pitch perception.

Pitch is a musical term that refers to the lowest frequency we perceive in a tone. This lowest frequency is the *fundamental frequency*, or simply the *fundamental*. In a *simple tone*—a sound that contains only one frequency—the pitch is identical to the fundamental frequency. Notably, simple tones are rare in real-world music and are typically studied in audiology rather than musical contexts. Most music consists of *complex tones* containing the fundamental frequency and a series of *harmonic frequencies*. The harmonics are integer multiples of the fundamental. Our brains use this structured pattern of harmonics to determine the pitch of a tone rather than relying solely on the fundamental frequency itself. (There are three fascinating observations regarding human pitch perception.)

1. The Range of Pitch Perception

Our ability to perceive the pitch of a complex tone is limited to a specific frequency range—approximately 30 Hz to 5,000 Hz, as presented in the paper "Pitch as a Medium: A New Approach to Psychophysical Scaling" by F. Attneave and R. Olson. The consequence is that if a tune is played using tones above 5000 Hz, we no longer perceive it as a coherent melody. Interestingly, this range aligns closely with that of a grand piano, suggesting that our perception of pitch is well-suited to the instruments we have developed. Even the highest note found on any standard musical instrument—the C8 of the piccolo flute at 4186 Hz—falls within this range. Our musical traditions and instrument designs have evolved to fit within the natural boundaries of human pitch perception.

2. The Missing Fundamental

A fascinating aspect of pitch perception is that the fundamental frequency's physical presence is not required to be perceived. Instead, our brain reconstructs the fundamental by analyzing the harmonic series—a process similar to filling in missing information in a familiar pattern. Research has shown that even with as few as three consecutive harmonics, we can still perceive the missing fundamental. Even more remarkably, these harmonics don't need to be the lowest in the sequence (e.g., the 2nd, 3rd, and 4th harmonics); they can be much higher, such as the 23rd, 24th, and 25th harmonics. Notably, as the pitch increases, the harmonics required for this effect must be lower in the sequence.

This phenomenon explains how our brain compensates for dips in our stereo system's frequency response, particularly in bass frequencies. If a sound system lacks strong low-frequency output, our brain still perceives the fundamental pitch as long as its harmonics are present. Although, while we may still perceive the fundamental tone, the sound will feel less satisfying than having the fundamental frequency physically present. A good bass response in audio reproduction is important for a fuller, more immersive listening experience.

3. Non-Linearity in Hearing: The Tartini Tone

Another intriguing non-linearity in our hearing is related to the Tartini tone, a phenomenon first documented in 1754 by the composer and violinist Giuseppe Tartini. This effect occurs when two musicians play two different tones: in addition to the actual notes being played, we perceive an additional tone whose frequency is the difference between the two original frequencies. This "phantom" tone arises because our auditory system performs mathematical processing, effectively generating an extra pitch that isn't physically present. Musicians are well aware of this effect and often use it to their advantage. For instance, by carefully selecting intervals, they can create the illusion of a lower pitch than their instrument is physically capable of producing. This is particularly useful in small ensembles, where Tartini tones can enrich the harmonic texture of the music.

These three features—the range of pitch perception, the missing fundamental, and nonlinear pitch phenomena like the Tartini tone—highlight the remarkable nature of our auditory processing. Our brains don't merely receive sound passively; they actively interpret, reconstruct, and even create missing elements, fundamentally shaping how we perceive music and sound.

The Qualitative Nature of Tone

We readily accept that we can distinguish between a trumpet playing a C and a clarinet playing an F. This familiarity likely explains why tone is a significant topic in audio discussions. We know what a clarinet should sound like—its signature "woody" quality is a defining characteristic. Our music reproduction will not feel authentic if we fail to capture that.

This qualitative definition of tone, used by musicians and audiophiles, is more subjective and experiential. It refers not just to the presence of frequencies but to the character or quality of a sound—how it feels and is perceived by the listener. Instead of numerical descriptions, qualitative tone is often expressed through descriptive terms such as warm, bright, harsh, smooth, brassy, nasal, or airy. These descriptors capture how a sound interacts with the human ear and emotions, making tone a critical element in musical expression and audio reproduction.

A musician's ability to shape and control tone using their lips is particularly important in wind and brass instruments. Even subtle changes in embouchure—how the lips, facial muscles, and tongue interact with the instrument—can dramatically alter the sound. This control allows musicians to create a wide range of tonal colors, from bright and piercing to warm and mellow. The ability to shape tone with the lips is not just about achieving a "good" sound—it's an essential tool for artistic expression. Musicians develop their embouchure technique over years of practice, refining their ability to adapt their tone to different styles, moods, and dynamics. Whether producing a lush, velvety legato or a sharp, biting attack, the lips serve as a critical interface between the musician and their instrument, allowing for an extraordinary level of nuance and control over their sound.

Just as musicians shape the tone of their instruments through technique and physical control, high-fidelity (HiFi) audio systems also exhibit distinct tonal characteristics based on how they are set up and fine-tuned. In the audiophile world, tone refers to the overall sonic character of a system—how it reproduces sound in terms of warmth, brightness, clarity, and balance. For example, a system that emphasizes higher frequencies might be described as bright or edgy, while one that enhances lower frequencies might be considered warm or woolly. These tonal attributes are not fixed; Subtle changes, such as angling speakers differently, adjusting their height, or modifying the listening environment with acoustic treatments, can significantly alter the perceived tonal balance. The musician and the audiophile rely on a mix of technical understanding and a trained ear to achieve the most pleasing and expressive sound possible.

Temporal and Spatial Resolution

Temporal resolution refers to the ability of our hearing to subdivide time and distinguish individual sounds with fine precision. A key question in music perception is: How much time resolution is required to distinguish each note individually? Performing a simple calculation based on a well-known piece of music gives a first-order estimate.

Consider the third movement from Vivaldi's "The Four Seasons," Summer, marked presto—a tempo ranging from 168 to 177 beats per minute (BPM). The piece is written in ¾ time, giving a quarter note one beat and having 3 beats per measure. The violin plays sixteenth notes, with each note lasting one-quarter of a beat. Setting the tempo at 170 BPM translates to 2.8 beats per second (170 beats ÷ 60 seconds). A quarter note, one beat long, lasts 357 milliseconds (1000 ms ÷ 2.8). Finally, the sixteenth note, one-quarter of a beat, would last 89 milliseconds.

Since each played sixteenth note is clearly distinguishable, we establish that our auditory system has a minimum of 90 milliseconds of temporal resolution. As we will see, our ears are capable of much finer resolution.

Musical note spacing is one of two key aspects of our hearing's time resolution. Audiologists interested in the smallest time gap detectable by the human ear devised testing methodologies. Results from early experiments, indicating single-digit microsecond time resolution, introduced a brief silence within a continuous tone. The experiments were flawed, and the results were incorrect. The abrupt stopping and restarting of the tone introduced frequency-related artifacts—additional high and low-frequency components—that influenced perception. Researchers suspected listeners detected these spectral cues rather than the actual time gap. (One could argue that it does not matter how someone detects the time gap. It is still detected.) More sophisticated tests were developed using white noise masking to eliminate frequency-based artifacts, establishing the temporal resolution of human hearing remains in the low single-digit millisecond range.

As demonstrated in our calculation from *The Four Seasons*, our ability to perceive timing gaps far exceeds what is necessary to distinguish even the fastest musical passages. This exceptional time resolution is essential for music appreciation—it also plays a critical role in speaker setup and audio system optimization, enabling us to optimize for a more accurate listening experience.

A second aspect of time resolution is our ability to detect differences in sound arrival time between our two ears, which is crucial for determining sound direction. In 1907, Lord Rayleigh proposed the *duplex theory* in his work *On Our Perception of Sound Direction*. This theory explains that humans localize sound in the horizontal plane using two primary cues: The Interaural Time Difference (ITD) — the slight delay between when a sound reaches each ear. The Interaural Level Difference (ILD) is the difference in sound intensity between the two ears. Together, these mechanisms enable precise spatial localization of sound. The duplex theory remains widely accepted today and has not been refuted.

Alan Blumlein expanded on Lord Rayleigh's work and played a pivotal role in developing the principles of stereophonic sound. It is remarkable that the patent for stereo sound dates back to 1931. To understand more, let's turn to one of the most influential studies in audio science.

In 1933, Blumlein published patent 394325 in the United Kingdom, marking the culmination of years of research on stereophonic audio. In the same year, a musical concert by the Philadelphia Orchestra was transmitted from Philadelphia to Washington, where a stereo array positioned in front of the audience was used for its reproduction. This event marked the realization of Blumlein's dream: allowing those not present at a symphony to experience the music as though they were truly there.

Although some work on stereophonic sound existed before Blumlein, his groundbreaking research refined Lord Rayleigh's theories on sound localization and formally established the principles still used today. In his patent, he stated: "Directional hearing sense is due to phase and intensity differences between sounds reaching the two ears, phase differences being more effective for the lower frequencies and intensity differences for higher frequencies." His work laid the foundation for modern stereo recording and playback.

Lord Rayleigh's and Blumlein's insight into the two core mechanisms our brains rely on for locating sound: interaural time differences and interaural level differences, are fundamentally important. Understanding how these cues work—and where they fall short—is essential for grasping how stereo imaging is created and why precision speaker placement matters.

Interaural Time Difference (ITD) and Interaural Level Difference (ILD) function because the width of our head separates our ears. These two mechanisms allow us to determine the location of a sound in the horizontal plane, as depicted here, but their effectiveness depends on frequency.

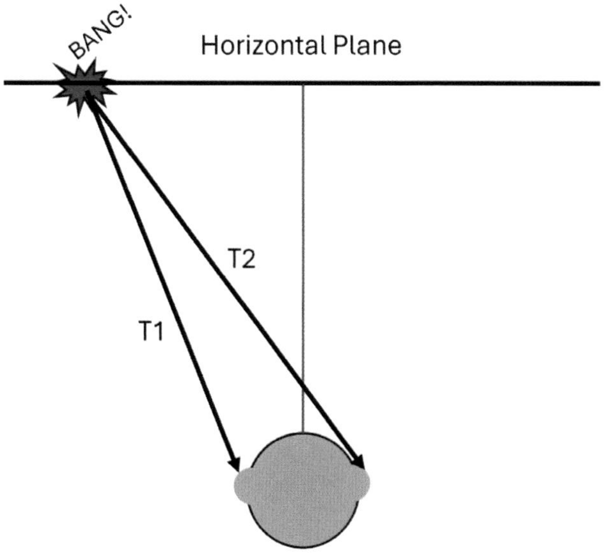

Before exploring how timing and level cues shape stereo imaging, we must understand how these mechanisms behave across different frequencies.

For lower frequencies (below approximately 1.5 — 1.7 kHz), ITD is the dominant method of localization. In this range, a sound wave reaches one ear slightly before the other, creating a brief time delay. Although, as frequency increases, this method becomes less reliable. At higher frequencies, additional wave crests arrive at the first ear before the second ear detects the original crest, causing ambiguity in determining which crest arrived first. This effect sets an upper-frequency limit for ITD-based localization.

Above this limit, ILD becomes the primary localization mechanism. At higher frequencies, where wavelengths are shorter than the head's width, the head acts as a sound barrier, creating a shadow effect. Instead of diffracting around the head, high-frequency waves are reflected, causing a significant drop in loudness for the ear opposite the sound source.

By combining these two mechanisms, the brain constructs a spatial map, associating specific time delays and loudness differences with the horizontal location of a sound source.

Research indicates that humans can localize sound sources with a horizontal spatial resolution of approximately 1 degree. This means that within a 60-degree arc—such as the angle formed by an equilateral triangle between a listener and two speakers—there are about 60 distinct positions where a sound can be perceived. This level of precision is evident in audio test tracks, such as the one found on IsoTek's Ultimate System Setup disc (track 6), which features a wood block moving incrementally from left to right between the speakers. The track contains 33 clicks, allowing listeners to easily discern each discrete position of the block, which confirms our ability to detect fine spatial changes in sound source placement.

In addition to interaural time differences (ITD) and interaural level differences (ILD), our auditory system employs a third mechanism for sound localization, particularly in the vertical plane. This involves the unique shape of our outer ears, or pinnae, interacting with incoming sound waves. When high-frequency sounds, typically ranging from 5 kHz to 12 kHz, reach the ear, they are reflected and diffracted by the complex structures of the pinnae. These interactions modify the sound's amplitude and phase before it enters the ear canal, creating distinct spectral patterns that our brain interprets to determine the elevation of the sound source.

In the context of audio playback, the orientation of speakers can influence these vertical localization cues. Adjusting the rake angle—the tilt of the speaker either forward or backward—affects how high-frequency sounds are directed toward the listener. Tilting the speakers can change the angle at which these frequencies arrive at the ears, thereby altering the perceived height of the sound source.

Sound localization depends on our brain's ability to accurately assess the time differences between the ears. Understanding how accurately we can hear time differences will give us insight into how precisely we can expect to adjust speaker placement. This was the subject of a 2019 paper by Thavam and Dietz titled "*Smallest perceivable interaural time differences.*" The research highlighted several interesting observations:

- "It is still noteworthy that thresholds for untrained listeners are 2.6 times higher than for trained listeners."
- "The average threshold ITD across nine trained listeners in a two-alternative forced-choice task is 6.9 microseconds."
- "Two of the nine listeners performed significantly above chance at an ITD as small as 3.7 microseconds."
- "For 52 untrained listeners, the average threshold ITD is 18.1 microseconds."

There are two important points. First, the capability of human hearing to comprehend differences of less than 5 microseconds. Second, our ability to train our listening skills to hear finer and finer timing differences. Both of these are important in speaker setup. Do not underestimate your ability to hear very fine details and timing differences. As we train ourselves to become better listeners, we can make better decisions about the necessary speaker corrections to resolve timing errors.

Stereophonic Sound and ITD/ILD from Two Sound Sources

Sounds we experience in everyday life are naturally coherent and emanate from a single point in space. Whether that is the chirping of a bird or the sound of a construction worker's jackhammer, the sound's harmonic structure is preserved, and our brain processes the ITD and ILD information, providing us with the sound's location. Conversely, stereophonic sound is a contrived mechanism that creates the illusion of an object existing at a specific point in space. This object is called a **phantom image**. The stereophonic presentation of all phantom images is referred to as the **soundstage**.

The phantom image's position within the sound field is determined by the amplitude and phase (time) information gathered by the stereo microphones during the recording process. If this is a live recording, each musician's position is represented by one phantom image. It is also possible and common to manipulate the time and amplitude information during the mixing process, where sound engineers adjust the signal's relative right-to-left time delay or amplitude. The engineer will reference a particular set of "panning curves" that contain the exact time delay or amplitude adjustment required to specify the phantom image's precise position within the soundstage. For example, using a panning curve, a time delay of 1 millisecond to one of the channels will shift the image to the far left or right. Additionally, changing the amplitude (loudness) of one channel relative to the other by 15 dB will shift the phantom image to the far left or right.

Recall the conservative estimate that humans can discern 1 degree of lateral spacing between sounds, allowing for 60 distinct positions between the loudspeakers. If a phantom image can be pulled far to the left by a 1-millisecond delay, then each degree corresponds to approximately 33 microseconds of delay.

Creating the correct time and spatial relationships captured on the recording is the challenge of speaker placement. An ideal speaker setup would have each speaker perfectly coherent and both speakers coherent with each other. Once this is achieved, the overall system will emulate a natural point source sound. Loudspeakers present a more complex situation as they don't have a single ITD and a single ILD, like a sound in nature; they have an additional time and level delay called the inter-channel time difference (ICTD) and an inter-channel level difference (ICLD).

The next diagram illustrates two loudspeakers positioned symmetrically in front of a human head. Notice the lines labeled T1 and T2 representing the time delay between the speaker and the nearest ear. Also, note that there are two additional lines labeled T3 and T4, representing the time it takes for the signal from each loudspeaker to reach the farther ear. The ICTD, T4-T2 or T3-T1, can be calculated and depends on the width of the head, the angle the listener's ear makes with the forward direction, and the angle between the loudspeakers. For a typical head with a radius of 9 cm and a 60-degree angle between the speakers, the time delay would be approximately 0.27 ms. This is relevant as De Sena showed that moving the head laterally by more than this amount resulted in localization uncertainty. This gives the width of the stereo "sweet spot" as approximately 3 head widths.

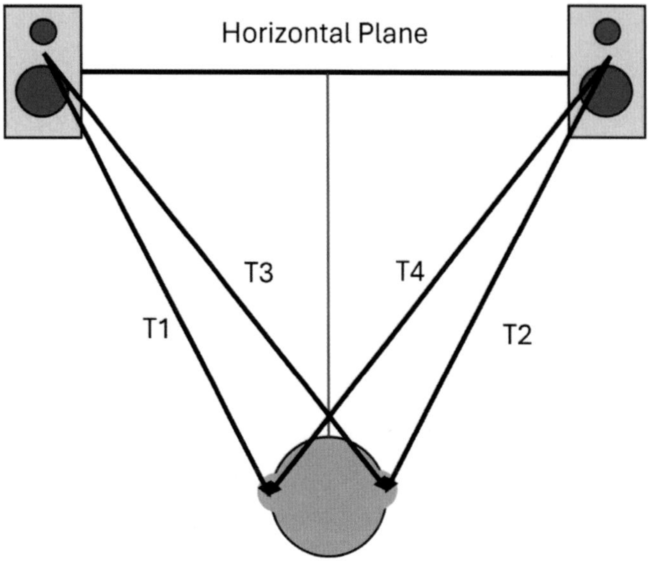

The ICTD and ICLD of two sound sources, as presented by the two loudspeakers, cannot be treated as identical to the ITD/ILD of a single point source. As Enzo De Sena points out in his paper, *Localization Uncertainty in Time-Amplitude Stereophonic Reproduction*, "A common confusion in this context is to assume that ICTDs are identical to ITDs, and ICLDs are identical to ILDs, which is incorrect. The wavefronts of each loudspeaker reach both ears and form interference patterns at the position of the ears, which, in turn, lead to a complex relationship between ICTD-ICLD and ITD-ILD. Also notice that, as opposed to ICTDs and ICLDs, ITDs and ILDs are frequency-dependent."

Again, considering the diagram above, it is clear that for each sound event produced by the speakers, each ear receives two impulses — one from the near speaker and one from the far speaker. Audiologists categorize these events into one of two types: short impulses and sinusoidal. Musically, examples of these might include rim shots for short impulses, and male vocals would closely approximate sinusoidal signals. We will focus on the short impulses and comment on the sinusoidal observations.

When the speakers are positioned perfectly symmetrically, a short impulse played at the same amplitude from both speakers reaches both ears at the same time—twice. ***(Left illustration.)*** In this case, the phantom image forms precisely in the center between the speakers.

Granted, if the right speaker is positioned even slightly farther from the listener, the signal from that speaker will arrive at the ears with a small delay. ***(Right illustration.)*** In this situation, the left ear first receives the sound from the left speaker, followed by the right ear receiving the sound from the right speaker. Shortly after, the right ear also receives the delayed sound from the left speaker, and the left ear receives the delayed sound from the right speaker.

This sequence of arrivals creates a subtle timing imbalance that shifts the phantom image toward the left. One might assume the image simply moves toward the side that hears the first arrival, but studies by Toole and Sayers (1965) revealed that image localization under these conditions is far more complex, depending on the interaction of timing, phase, and amplitude cues at both ears.

Quoting from Blauert, "One experiment conducted by Toole and Sayers in 1965 showed — one click A was presented to the left ear and two clicks B and C were presented to the right ear. The subjects could identify two components to the phantom image. One was a "low-pitched" component whose path was determined by the positions of the A and B impulses relative to C and a "high-pitched" component whose lateral displacement was determined only by the relative position of the A and C impulses." What can be concluded is that, with impulse signals, both the initial signal and later-arriving signals at the ears contribute to phantom image localization.

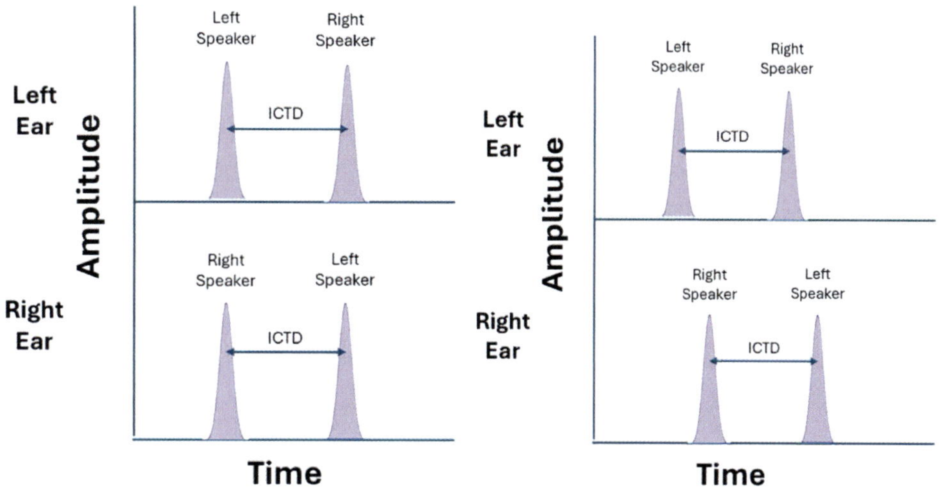

Examining the illustration above, it is evident that the lowest noise will occur when the speakers are perfectly aligned. This is the case where we experience a single time delay. Any misalignment of the speakers will result in two separate time delays instead of hearing dink … dink. We will hear d-dink ... d-dink. One interesting observation from speaker setup is that as the speakers become closer and closer to perfect alignment, they will start to sound bright. I believe this is because the distance between the two impulses becomes increasingly shorter, producing a high-frequency sound that our ears can perceive.

Sinusoidal signals, approximated by vocal tracks, behave very differently than short tone bursts. Two important observations regarding the behavior of sinusoidal signals are noteworthy. The first key observation is that at low frequencies, time differences between speakers result in level differences between the ears. In contrast, amplitude differences between speakers result in time differences at the ears. The second observation is that sinusoidal signals are only effective for phantom image localization up to approximately 800 Hz (as noted by Blauert).

Utilizing vocal tracks for speaker positioning will turn positional offsets between the speakers into level differences. This is a different way of assessing the correctness of speaker placement. This can highlight problems with the position of the right speaker, as we may experience inconsistencies between what the ICTD and ICLD are telling our brain. De Seta points out that the conflict between ICTD and ICLD will result in localization uncertainty and image ambiguity. But De Seta also points out that ITD seems to be more relevant than ILD in terms of lateral image shifting.

As we can see, the way the brain interprets input from two sound sources is a complex process. I think this quote from Blauert's book *Spatial Hearing* is insightful:

"The psychophysical relationships between the attributes of the ear input signals and the positions of auditory events are, however, as a rule not describable in terms of linear systems. In particular, the auditory event or events that appear when there is more than one sound source are not superpositions of the individual events that would appear in connection with each individual sound source. When there are multiple sound sources, additional laws and rules apply to the auditory event; the phenomena on which these laws and rules are based do not occur when there is only one sound source."

What Blauert is saying, in plain English, is that the diagram on page 20 and the calculations and experiments based on it are only a rough, first-order approximation of what happens when we listen to a pair of speakers.

Interaural time difference (ITD/ICTD) and interaural level difference (ILD/ICTD) are necessary for stereo imaging, helping us perceive sound location in space. When adjusting loudspeakers, even the slightest movements can significantly impact these auditory cues. For instance, shifting a speaker by just 1 millimeter introduces a time delay of approximately 3 microseconds, which is noticeable to trained listeners. This sensitivity underscores the importance of precise speaker placement. Even more minor adjustments, such as moving a speaker by 0.1 millimeters, achieved by a small bump or tap, will produce audible and repeatable changes in sound perception. While this level of precision may seem surprising, it is demonstrable and highlights the intricate relationship between speaker positioning and our auditory system's ability to detect subtle temporal differences.

When positioning the right speaker, we aim to place sound sources accurately in space and time. For instance, a human voice on a vocal track will typically appear perfectly centered between the speakers. Earlier data was presented suggesting that humans have a spatial resolution of one degree. However, I propose that our resolution is even finer, especially for centrally located sounds. Consider a single sound source, such as a wood block struck ten feet away, versus a pair of speakers reproducing that sound. A real object directly ahead would produce zero time and amplitude differences between our ears. When using two speakers to simulate this, they must create an almost negligible time delay to make the sound appear centered. Our auditory system is highly sensitive to these time differences. When playing a song where the singer is mixed to be centered, I can detect an off-center image by less than an inch, even with speakers spaced 102 inches (8.5 feet) apart. This suggests a resolution of at least 0.25 degrees for centered sounds. This heightened sensitivity may be due to the sustained nature of vocals, which allows the auditory system multiple opportunities to localize the source with each word sung.

Summary: What We Hear and Why It Matters

I regret to inform you, dear reader, that the science of human hearing remains incomplete. It cannot fully account for the profound impact that small changes in speaker position have on perceived sound. That said, existing research still offers valuable insights. The general findings in the literature align with many of my own observations regarding speaker positioning.

The key discrepancy lies in scale. Audiologists typically study timing differences in the millisecond or hundreds-of-microseconds range. Yet in precision loudspeaker placement, audible and predictable changes can occur with movements corresponding to single microseconds—an order of magnitude finer than most research addresses.

This chapter has offered a basic overview of hearing as it relates to stereo setup. It is not definitive—entire books are devoted to this complex subject. The references cited at the end of this book provide a starting point for deeper exploration. *Still, we hear what we hear.* One may hold a PhD in audiology and master every current theory yet still lack the practical skill to position a pair of speakers so that music sounds truly convincing.

What we can say with confidence is that hearing is not a passive process. The brain is constantly analyzing timing, frequency, and spatial relationships to determine where sounds originate and how they relate to one another.

Two key mechanisms play an important role in this process: **interaural time differences** and **interaural level differences**. These cues allow the brain to determine the direction and distance of sound sources by comparing the tiny differences in arrival time and intensity between the two ears. Even differences measured in microseconds can influence our perception of location and clarity.

The auditory system also places tremendous importance on transients and timing cues. The initial onset of a sound often carries the most important spatial information. When these cues arrive cleanly and coherently, the brain can reconstruct a believable acoustic event. When they are blurred or delayed, the sense of realism begins to collapse.

Another crucial factor is **pattern recognition**. Over a lifetime of listening, the brain learns what voices, instruments, and acoustic spaces should sound like. It constantly compares what we hear to those internal references. This ability allows us to recognize instruments instantly and detect subtle deviations from natural sound.

These mechanisms help explain why small changes in a playback system—changes that may seem insignificant from a purely technical perspective—can dramatically alter our listening experience. The auditory system is extraordinarily sensitive to timing relationships and spatial coherence. When those relationships are preserved, music feels clear, stable, and engaging. When they are disturbed, the result can feel confusing or unnatural, even if we cannot immediately explain why.

Understanding how we hear provides the foundation for everything that follows in this book. But hearing alone does not explain the full experience of music. The brain is not simply responding to isolated sounds. It is responding to **structure**.

Music is built from relationships between notes, phrases, and instruments. Timing, harmony, rhythm, and orchestration all work together to create patterns that unfold over time. These patterns allow us to follow a melody, anticipate rhythmic movement, and feel tension resolve in a chord progression.

In the next chapter, we will explore how these musical structures are organized and why they are essential to the listening experience. By understanding the internal logic of music itself, we will begin to see why accurate reproduction of those relationships is so critical.

In short, before we can understand how to reproduce music faithfully, we must first understand how the human ear and brain interpret sound.

CHAPTER 2

MUSIC, STRUCTURE, AND THE LOGIC OF SETUP

Before diving into technique, it helps to recognize that music—and therefore sound reproduction—is not random. Both follow structures that define how beauty emerges from discipline. To understand this, let's borrow two worlds where structure defines excellence: classical music and baseball.

Music has structure, and in the same way, so does this speaker setup process. Rondo, Minuet & Trio, and Passacaglia are a few thematic processes a composer can use to organize music in a specific and recognizable way. Composing within a thematic structure requires discipline to make the music understandable. Most fundamentally, these thematic processes provide structure for organizing time. For instance, the rigid Passacaglia structure starts with the foundational theme, a bass line that cycles continually throughout the music. Listeners familiar with musical structures will recognize the structural theme soon after the music begins. It's like acknowledging that now we're watching a baseball game, not a football game. The shared knowledge between the composer, musicians, and audience regarding the overall Passacaglia structure provides an expected framework, allowing artistry to emerge and be appreciated. What is not known is where the composer will take the music within that structure. The Passacaglia bass line cycles allow a composer to add melodic variety over the stable bass foundation, creating a complete musical work.

The better one understands the Passacaglia structure, the more one recognizes the beauty of the nuanced variations relating to the bass line. As it turns out, some compositions are masterpieces with elegant, playful, or dire variations, bringing you on a sophisticated journey. Other compositions are just pretty music. Truly understanding the structure reveals the composer's genius and allows the listener to understand how the composer captivates the audience. The Passacaglia's casual listener is likely to remain unaware of its underlying complexity and, therefore, is less appreciative of it. That doesn't mean the music won't deeply move them, but they probably won't understand the beautiful path that brought them there, and they ultimately have a less enriched experience. Not understanding the Passacaglia structure doesn't ruin the ride if one prefers to enjoy the experience rather than study the journey. A non-musical analogy makes this structure even easier to visualize. And to illustrate this, let's consider baseball.

To a first-time observer, baseball is simply a game with a winner and a loser. They are unfamiliar with the structure, rules, and nuances. Perhaps the casual observer understands some of baseball's structure: nine innings, three outs, three strikes, and four balls. The casual observer can find the game's final score in the news and have some information about what happened. For example, a score of 2-0, could tell us who won and that it was generally a defensive struggle. A score of 11-10 in extra innings tells us who won the offensive slugfest where one team finally outhit the other in the end.

The slugfest might be far more fun to watch to the casual observer than the defensive struggle. But what if the 2-0 game was a "no-hitter?" That's the pitching equivalent of a great composer's masterwork. The descriptive "no-hitter" changes everything. As a further nuance, let's examine the final pitch of that 2-0 no-hitter. It's the top of the ninth inning with one out. The pitcher walked the second batter, who is now on first base. The stage is set with the third batter up, a specially chosen pinch hitter who is a fast runner. On the first pitch, the batter hits a looping shot to shallow left-center field. The left fielder trips on his way to the ball, but the center fielder is there and *barehands* the ball after the first bounce. He's right-handed but still manages to twist and flick the ball to the second baseman, who grabs it in full flight, tags the runner out and in mid-air, clotheslines the ball to the first baseman, just beating the guy with serious wheels to the base, retiring the inning and the game, and clinching the no-hitter. Any potential no-hitter is nine innings of building suspense. Still, this one culminates in a difficult-to-execute but exquisitely played double play and is the musical equivalent of a Beethoven finale. Each pitch of the game was its own microcosm within baseball's structure, encompassing each out, each inning, and ultimately, the heart-stopping game finale.

Whether in a concert hall or a ballpark, mastery lives in the relationship between **precision and freedom**—and loudspeaker setup is no different. This structured unpredictability has a direct parallel in loudspeaker setup. So how does this have anything to do with speaker placement?

The answer is consequences. In baseball, anything can happen with each pitch, and each pitch impacts the game's conclusion. With the speaker set up, every miniscule positional change affects the results. Although, it is far from random, and like the Passacaglia, a process can have a rigid structure yet allow for variations in execution to achieve a beautiful result. Recognizing how each setup process step relates to the foundation allows freedom to adjust to each individual system, speaker type, and room interaction we will encounter. A wrong move at the end of a great speaker setup can ruin the "perfect game". But perfection isn't random. We can fully appreciate a composer's detailed score with a stereo system fully optimized. We can hear all the emotions the musicians pour into the performance.

The Goal of Speaker Positioning

Our home stereo is there to reproduce music. Thus, if our stereo is to produce music, we should ask what is required for music. Often, audiophiles use terms like detail, soundstage, imaging, warmth, etc., to describe what they hear. These are descriptors of sound, and they are mere passengers when speakers are correctly set up. We want something more. We want real music performed by real musicians. It turns out that music only requires two things, with a third that adds artistic touch. These are tones, rhythm/timing, and dynamics (loudness variation). Every change to the stereo system affects one or more of these three things. What are they, and how do we persuade our stereo to faithfully reproduce them?

Tone: The Foundation of Musical Expression

The first thing required for music is a range of tones. These are the sounds made by instruments playing notes placed on the music staff, denoted C, D, E, F, G, A, and B. Every note played has a specific pitch (fundamental frequency) and timbre (harmonic series). Together, pitch and timbre form a series of distinct, unique frequencies—think of it as a sound fingerprint. With the wide array of musical instruments and voices, this paints a sonic picture with vivid color. Collectively, this is tone. For classical music fans, think Claude Debussy.

How do we get tone? More importantly, how do we get the **correct** tone? The answer is **speaker placement.** If the speaker system is improperly balanced, it will negatively affect an instrument's tone. For example, imagine listening to a Bach concerto; a trumpet is playing C5 at 523 Hz. Let's assume the speaker system creates an interaction (such as a reflection) that causes a sharp increase in our frequency response at 1046 Hz (the second harmonic of 523 Hz), making it 4 dB louder. Every time the trumpet plays C5 at 523 Hz, the second harmonic becomes too loud relative to the fundamental frequency. This imbalance distorts the trumpet's timbre, making it sound unnatural. We will still recognize it as a trumpet, but there will be some unease as it will be a much less accurate rendition of a real trumpet. Proper speaker placement hopes to remedy this tonal anomaly. Our speaker placement goal is to **smooth and balance** the tonal response as much as possible. To adjust the system's tone, we make relatively **large speaker movements**. Finding the correct **front-to-back**, **lateral**, **height**, and **toe-in** is about balancing the tonal response between the speaker and the room.

Rhythm: The Pulse That Brings Music to Life

With tone established, the next foundational element is the organization of musical time. The second requirement for music is rhythm. Rhythm is the mechanism that allows instruments to play together in time. If musicians played their parts whenever they wanted and at any speed, we would hear a cacophony rather than a symphony. The sound would be a random mass of notes. Instead, rhythm and timing infuse life into the various tones. For classical music fans, think J.S. Bach.

How do we get the musicians to play together and achieve a sense of rhythm?

Again, adjustments to the speaker position are the answer. This often gets overlooked in speaker setup. Large adjustments to the speaker position will correct **tone**; smaller adjustments help **fine-tune timing** (and further enhance tone). You might be surprised to learn that for coarsely aligned speakers, the drivers' output doesn't reach your ear at precisely the same time. We need to time-align the **left speaker's woofer with the midrange**, and the **midrange with the tweeter**. Then, we must synchronize the **right speaker with the left**. Once this is completed, the sound will reveal the musicians in the band or orchestra playing together. The music will then begin to take on meaning.

Dynamics: The Soul of Musical Expression

The last thing music needs is the breath of life—emotion. Music is imbued with emotion when we feel the ebb and flow, the ups and downs, and the excitement and despair. Dynamics, or the variation of loudness, is responsible for this. Each instrument can play as softly as possible, marked as ppp on a music score. The instruments can play as loudly as possible, marked fff on a music score. This variation in loudness lends the music interest and enables the musician to express their artistic interpretation. For classical music fans, think Ludwig van Beethoven.

If dynamics represent the soul of music, then how do we maximize it? Increasing dynamics can be achieved in two ways. First, **lowering the noise floor** of the overall system will allow us to hear quieter sounds. This increases the perceived distance between ppp and fff, thereby increasing the dynamic range. Second, **time-aligning the speakers** reduces time smearing, allowing us to hear the artist's intent more clearly. Time alignment produces more space between notes, with the start of notes arriving at the ear at precisely the same instant. The impulse response's fast rise time conveys quickness. The space between notes allows the decay after the note to happen naturally without truncation. Allowing this space between notes **lowers the perceived noise floor**, thereby increasing the sense of dynamic expressiveness. We need to remove as much noise as possible, either acoustic or electronic, to achieve the musical connection we are looking for.

The Integration of Elements

Once these three elements are understood individually, their true power emerges when they function together. Proper speaker positioning integrates **tone**, **rhythm**, and **dynamics** into a seamless musical experience. **Large** positioning adjustments establish proper tonal balance between speakers and the room. **Fine** adjustments achieve precise timing alignment. Together, these create the low-noise, high-resolution platform that allows dynamic expression to flourish. When executed properly, your speakers disappear, leaving only the music. You'll hear real instruments played by real musicians in real acoustic space—the ultimate goal of high-fidelity reproduction.

These principles become unmistakable when you hear them through trained ears.

What Musicians Hear: A True Story

Recently, while working on a system for two professional musicians from the San Francisco Symphony, I asked them to comment on the music during the final aspects of the set-up process. I knew where I was in my process, but I was curious to hear their reactions as I progressed. I was not seeking feedback or help on what to do; I was interested in how they perceived the changes as they occurred. They had several observations: "The orchestra sounds rushed. They sound nervous." Then, at one point, a comment, "One of the violas is sloppy with the changeovers." Another comment was, "I hate that old-school squawky oboe style." And finally, "The violins are not ringing." These observations all relate to **tone** and **timing**. While adjusting the speakers, the tone and timing steadily improved. "That sloppy switch-over from the viola went away." "The oboe tone is fuller, and almost all of that squawky-ness is gone." "The orchestra has settled down and sounds controlled." By the end of the setup, the client commented, "That orchestra is much better than I thought." The initially nervous, sloppy, squawky orchestra played beautifully—not because the recording had changed, but because the speaker setup allowed the speakers to reveal the actual beauty in the recording itself. This transition occurred through very careful and precise speaker positioning.

In the end, a perfectly tuned system behaves like a great orchestra under a master conductor—structured, disciplined, and alive with expression. That's what high fidelity truly means.

Before we practice the precision that makes this possible, let's clear away ten common myths that reliably steer listeners off course.

Ten Myths and Misconceptions About Speaker Placement

Of course, many people have experimented with speaker placement over the past century since the stereo was first invented. In some cases, conclusions have been drawn that are not correct. These "facts" have been passed down and have become part of audiophile knowledge; read them carefully and open your mind to the possibility that what is considered fact is actually fiction.

1. **Speakers as Coherent Sound Sources** — It is commonly believed that speakers function as perfectly coherent sound sources. While manufacturers often emphasize phase coherence at crossover points, achieving true coherence at the listening position is more complex. In an actual room, sound from the speakers reflects off boundaries such as walls, floors, and ceilings, causing specific frequencies to arrive at the listener's ears either in or out of phase. This leads to phase cancellation or reinforcement, affecting the overall sound quality. Proper speaker placement aims to optimize phase coherence across all frequencies at the primary listening position and throughout the room.

2. **Relationship Between Toe-in and Brightness** — Many people firmly believe that toeing-in speakers causes brightness. Many have experienced increased brightness or harshness when adjusting speaker angles inward, concluding that toe-in is the cause. However, this perception often arises from being outside the tweeter's optimal dispersion window. As speakers are toed in, the high-frequency energy directed toward the listener increases, which can lead to a perception of brightness if not properly aligned. Finding the correct toe-in angle where the tweeter's response is most linear is crucial. This process involves careful experimentation, as each speaker and room combination is unique. Some speakers, for instance, have design features like a small covering in front of the tweeter, altering dispersion characteristics and phase relationships. Such designs may require less toe-in to achieve optimal sound. Therefore, it's essential to evaluate the appropriate toe-in for each speaker individually to achieve the best sound quality.

3. **Relationship Between Toe-in and Soundstage** — Another misconception about toeing-in speakers is that a smaller soundstage will result. The toe-in angle will modulate the sound energy from the sidewall, with more toe-in reducing the amount of reflected sound. The ear will combine the direct and reflected sound, pulling the sound's perceived direction toward the wall, broadening the perceived sound field. Unfortunately, the result is a broad, unfocused, noisy sound. Ironically, many people put an acoustic absorber on the sidewall, which will reduce or eliminate the sidewall reflection, defeating the reason for toeing out the speaker. The speakers need the optimal toe-in position; an expansive soundstage will result when the setup is complete.

4. **Relationship Between Toe-in and Sweet Spot** — A common misconception is that pointing speakers straight ahead with minimal toe-in will create a wider sweet spot and a more uniform sound throughout the room. In reality, this approach will lead to a disorganized and chaotic soundstage. Without proper toe-in, sound disperses unevenly, causing listeners seated off-center to predominantly hear the nearest speaker, resulting in an imbalanced audio experience. Stereo systems are designed for optimal listening from a central position, achieved through coherent sound delivery. Proper speaker alignment ensures sound remains clear and balanced across a wide listening area and throughout the room.

5. **Necessity of Room Treatment** — Science has a strong pull and is where many audiophiles place their hope. Acoustic science dictates that extensive room treatment is necessary for optimal sound quality. From experience, only minimal treatment is typically required once the listening position and speakers are optimally arranged to work with the room's acoustics. The goal is to integrate the speakers with the room, allowing them to work together to produce clear and articulate sound. With proper setup, the bass will be tight, the midrange clear, and the highs smooth. For more detailed information, refer to the section on acoustic treatment.

6. **Speaker Distance from Walls**—Some believe speakers must be positioned far from the walls for optimal sound. Perhaps you have heard speakers require a distance greater than five feet from a wall to sound good. This is not true. The speakers can be optimized within two to three feet of the front wall

and still sound very good. Might the speakers sound better if they were moved 10 feet from the wall? Maybe? Likely? Everyone has their space and should optimize within that space.

7. **Relationship Between Tweeter Height and Soundstage** — Many speakers have a tweeter well above typical seated ear height. The belief is that a high tweeter will result in a tall soundstage. While it does achieve height, it often makes the performer sound as if they are singing from the ceiling. This has led to the mistaken belief that tilting speakers forward results in a shorter soundstage. When speakers are properly optimized, the soundstage naturally extends to the ceiling while maintaining the singer's appropriate height. I owned one pair of speakers where tilting them forward raised the soundstage. A tall soundstage can still be achieved even with speakers with low tweeters, such as bookshelf speakers.

8. **The Magnitude of Speaker Movements** — Many underestimate the magnitude of the change in sound compared with the distance a speaker moves. People often say, "I can't believe how much better the speakers sound after moving them just an inch." In speaker setup, an inch is a significant distance. Large movements like this only occur in the initial placement stage. As fine-tuning progresses, adjustments become much smaller, often less than 1/64 of an inch. At this level, the movement is imperceptible to the eye, requiring recalibration of what constitutes a large or small adjustment.

9. **The Audibility of Small Speaker Movements** — Some believe that minute speaker adjustments are beyond human perception. While this may be true for an untrained ear, careful listening reveals how sensitive our hearing is. The human ear has astonishing abilities and can detect small differences. (Discussed in the chapter on sound perception.) Despite this, we tend to trust our eyes more than our ears. No one questions that a 4K or 8K screen is sharper than a 1080p display, so why doubt that we can hear extremely fine sonic details?

10. **Aural Memory** — Another common belief is that our aural memory is too short to judge subtle differences, and that expectation bias tricks us into perceiving improvements. While these factors are real, they must be considered in context. Speaker setup is not about comparing one new component to another—there's no inherent "better" or "worse" in the process, minimizing expectation bias. Instead, it is a methodical process, like sighting in a rifle scope.

    When calibrating a scope, a shot is fired and the result is observed. Then, based on experience, a precise adjustment is made, and the process is repeated until the target is hit. Speaker setup follows a similar approach: you listen, assess the sound, make a deliberate adjustment, and listen again. Unlike a rifle scope, there's no visible bullseye, but with training, one can establish small, objective sonic targets, gradually refining the setup until the desired result is achieved.

Various Methods of Speaker Placement

An online search for speaker placement methods yields numerous results. While people use various techniques to position a stereo pair of speakers, there is some consensus on what works best. Although, the effectiveness of these methods varies. In general, sound quality improves as placement accuracy increases. Let's explore some common approaches and assign them an arbitrary positional accuracy score from 1 to 10, with 1 being very poor and 10 being perfect.

The first method is placing speakers on the floor without much consideration—often on either side of a television, positioned close to a wall, and pointed straight ahead. This can be described as "random" placement, earning a positional accuracy score of zero. With this approach, the setup produces sound but lacks real stereo imaging. The result is a large, undefined sonic "blob" between the speakers. The bass tends to be boomy and lacks clarity, while the high frequencies are typically rolled off due to the misalignment of the tweeters, making the treble sound dull rather than detailed.

The next step is finding a position where the speakers have a more natural sound. This involves pulling the speakers forward from the wall a few inches at a time and listening to music. The speakers are shuffled forward as a pair to produce visual symmetry. At each position, the listener assesses whether they like the sound better. At some point, the listener becomes happier with how the speakers sound. This zonal placement method would have a positional accuracy score of 1 or 2, depending on how visually accurate the setup is and whether spikes or feet are used to level the speaker. The bass will be less boomy and more articulate in this setup process. Voices will sound more human. High-frequency information may start to appear. The sound in this method is still not well defined and remains quite "blobby." There is no timing accuracy.

An improvement on the previous "zonal" method is locating the speakers' positions using one of the formulas found online. Ratios of the room's length and width are typically involved—for example, placing the speaker one-fifth of the length and one-seventh of the width. The ratios produce different distances between the front and side walls, attempting to smooth the bass. Various ratios can be tried to find one that is "best." A listener reaching this level of speaker placement is likely aware of toe-in and will toe-in the speakers to some degree. The listener is usually unsure what to listen for to achieve the correct toe-in, so they aim the speakers at the shoulders or something similar. The bass will likely be better in this position, since that is what the listener was attempting to improve. The human voice (lower midrange) will probably improve due to toeing in the speakers. This listener may have the speakers on spikes or feet and adjust them so the speaker does not wobble. This is the first method that involves measuring, so it earns a positional accuracy score of 3. The sound using this method will have better bass smoothness, and the sound will start to gain focus. A vocalist's central image may start to form but is typically too large and pulled to one side.

Further improvement is possible with the previous two methods (zonal or formula) by increasing the measurement precision for the right speaker. Find and mark the center line of the seating area. The center of the listening chair is placed on this line. Once the position of the left speaker is found, the right speaker is measured from the center line to mirror the position of the left. The front and rear spike distances are measured to ensure the speakers have the same toe-in. The spikes or feet are adjusted using a bubble level to ensure the speaker is level in both the X and Y directions.

This improved zonal/ratio method would earn a positional accuracy score of 4, depending on how accurately the tape measure is read or if lasers are used. This method should produce a soundstage and a centered image, particularly with lower frequencies such as male vocals. It may not be perfect, but the sound is starting to develop from the "blob" we heard in the earlier methods. The bass will be as reasonable as in the zonal or formula-based system but will be more impactful because the speakers are better aligned with each other. High-frequency information is now appearing, which is what creates the soundstage. Understandably, the sound can be sibilant or harsh because of poor tweeter alignment. The sibilant, bright, or harsh sound often prompts the listener to start toeing the speakers out farther and farther until they are eventually outside the tweeter window.

I followed this exact path on my audiophile journey, spending increasing amounts of time on each step. Why didn't these methods work? There are a few reasons, but it comes down to positional accuracy. The left speaker has few positions where it works with the room, producing accurate bass, natural midrange, and extended, detailed treble. Formulas cannot be used to find them. These methods also don't address integrating the left speaker into a coherent, time-aligned source. Matching the right speaker is much more complex than it first appears. The floor is not level, and the walls are not square. Even if the left speaker were in a perfect position, obtaining a mirror image of the right speaker is impossible with simple measuring devices. In addition, subtle differences in the room's symmetry result in different interactions on opposing sides.

In general, the higher the level of positional accuracy, the higher the sound quality achieved. Of course, sound quality also depends on the specific position chosen within the room. Precision movements and adjustments to the speakers at any position will improve the sound of that position, producing the maximum sound quality attainable there. Granted, this may not be the best position available within the overall area in which the speakers can be placed. The goal is to find the best position—and then optimize around it.

Depiction of Sound Quality Versus Positional Accuracy

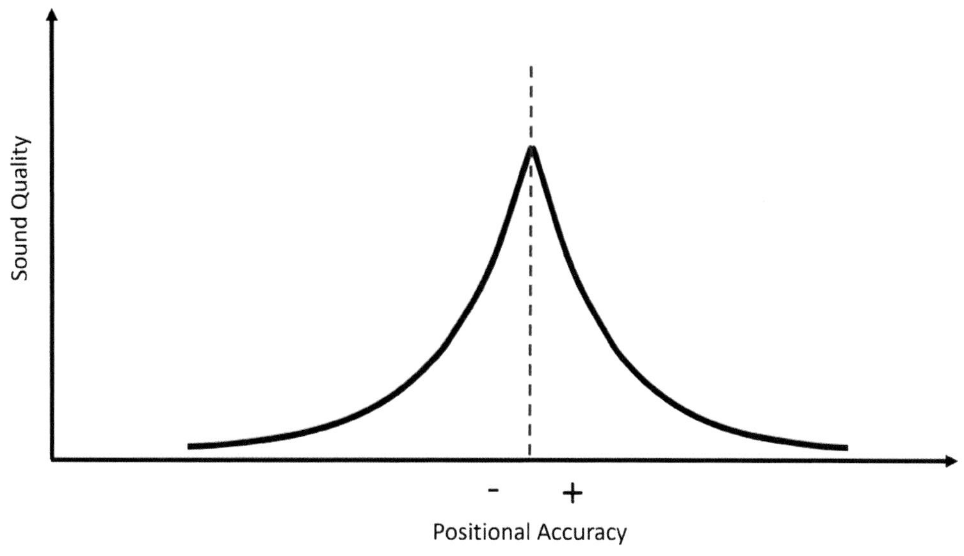

The Scientific Approach to Speaker Placement

One additional method of speaker placement I term the scientific approach to speaker placement uses a microphone and acoustic software such as REW to find the "right" position for the speakers. For technically minded people, acoustic measurements are the most obvious way to achieve good sound. I went down this path, taking thousands of measurements in my room. You can imagine how long this would take someone. Ultimately, this was an exercise in frustration because measurements could never tell me why my system could not produce satisfying music.

I remember marking my floor with a 28" x 28" grid of yellow masking tape, creating a 49-point reference for speaker placement. The grid's rear edge was about five feet from the front wall, leaving ample space between the speakers and the wall. I placed the front inside foot of each speaker at various grid intersections, carefully maintaining toe-in to ensure they pointed at the same spot. A clear pattern emerged through a series of acoustic measurements: moving left to right produced peaks and dips at specific frequencies, while moving front-to-back affected different frequencies. Using this data, I identified the position with the flattest overall frequency response and minimal compromise. Yet, despite my efforts, the speakers never truly "sang" or created what I considered real music.

After countless hours of frustration, I had the good fortune to meet Stirling Trayle—an encounter that changed the course of my audiophile journey. Upon seeing my meticulously taped grid with 4" x 4" spacing, I can only imagine his thoughts. Looking back, I find that grid both absurd and laughable. I had been misled by the overwhelming amount of "science" available for consumption. To be clear, I am not anti-science—I am a scientist. Notably, science is rooted in observation, and our primary tool for observing sound should be our ears. Stirling proved this in a single day. Relying solely on his ears, he found the speaker placement that allowed the room and speakers to work together to produce truly satisfying music. (For the record, that spot was not at the intersection of any yellow tape.)

Combining my measurement experience with the speaker setup techniques outlined in this book, I have come to a few conclusions. First, a flat and smooth frequency response in the bass range (20–100 Hz) is highly desirable. Second, a smooth RT60 response will produce excellent sound in the room. Third, the capability of acoustic measurements alone is insufficient to achieve the goal of true music reproduction. We will explore each of these.

A flat, smooth frequency response in the bass range (20 Hz to 100 Hz) is highly desirable—but not at the expense of coherence and transient response. Prominent peaks can smear both the bass and midrange, while dips are less intrusive—though once corrected, we often realize what we have been missing. Acoustic measurements are an efficient way to identify the ideal listening position, ensuring the smoothest bass response. We will explore this topic further later in the book. Measurements also help determine the best general placement for a subwoofer and ensure its phase is correctly set. Humans are relatively insensitive to bass frequencies, and without extensive training, it is difficult to distinguish between a dip at 50 Hz or 60 Hz or to gauge whether a peak is 4 dB or 6 dB louder. Measurements provide a quick and accurate way to assess bass response.

Second, while a smooth RT60 can contribute to good sound, it doesn't necessarily guarantee good music. RT60 measures the time it takes for an impulse to decay by 60 dB. Though there is debate about its relevance in "small" rooms (defined as anything smaller than an auditorium), I have found it often correlates well with our perception of good sound.

There's some confusion surrounding RT60, with some believing it's represented by a single number—like 0.5 seconds or 0.32 seconds. RT60 is not a single number but varies with frequency; some frequencies decay faster than others. The RT60 plot will also change as we adjust listening and speaker positions. What matters most isn't the exact average RT60 value (whether it's 0.7 or 0.5), but that the RT60 graph is as smooth as possible.

On the following page, I've included two examples of **RT60 plots**. In the first plot, there's a large peak at 300 Hz. Despite an average RT60 of less than 0.3 seconds, this resulted in remarkably poor sound—female vocals sounded thick and ringy. After adjusting the speakers, the second RT60 plot is much smoother. This improvement was achieved by making a series of very small adjustments to the speakers' positions.

RT60 Plot in Initial Position

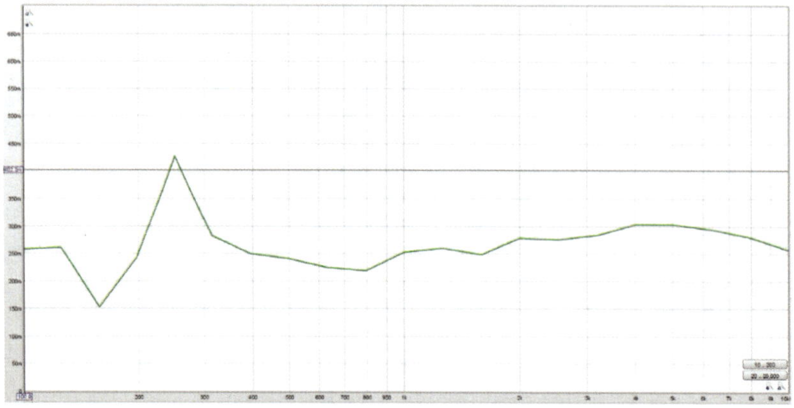

RT60 Plot in Final Position

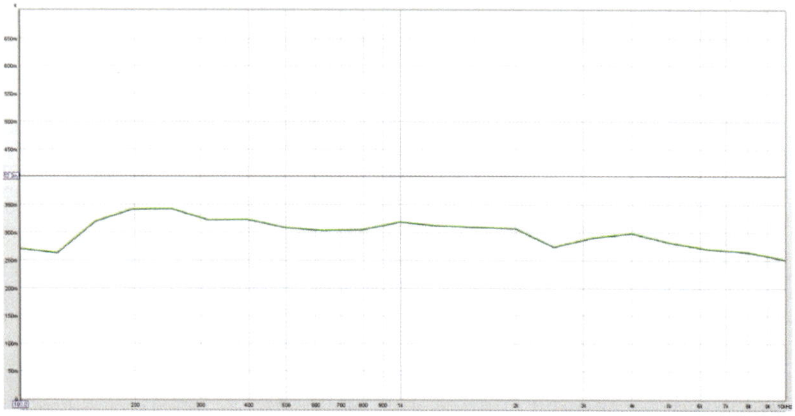

Following the steps in this book will help you find the speaker position in the room that provides a smooth frequency response and RT60. You might wonder—if that's our goal, why not rely entirely on measurements to achieve it?

The challenge is that our measurement tools aren't precise enough. While I'm unfamiliar with every acoustic measurement software package available, the options accessible to amateurs lack the accuracy required to perfectly time-align speakers to themselves and to each other. High-frequency data from measurements is highly sensitive to microphone placement. To assess left-to-right timing, the mic must be positioned exactly at the center of the listening position—where even a 1/16" shift can distort the reading. However, when adjusting by ear, we're making movements far finer than that.

Another limitation of impulse response measurements is that they only accurately reflect the distance to each tweeter. For example, the right speaker could be closer to the center but positioned farther back, yielding the same measured distance to the tweeter as a speaker placed wider and farther forward. Relying solely on measurements for speaker alignment can therefore give misleading results.

Acoustic measurements are a valuable tool that can provide useful information. Learning to take basic measurements and applying this knowledge can be helpful. But once you've honed your listening skills and you understand the adjustments needed based on what you hear, the process becomes much faster and more accurate than relying on acoustic measurements.

In the end, a perfectly tuned system behaves like a great orchestra under a master conductor—structured, disciplined, and alive with expression.

Summary: When Musical Structure Breaks Down

In this chapter, we examined how music is constructed—how rhythm, harmony, and orchestration create relationships that our auditory system recognizes and follows. These relationships give music coherence. They allow us to perceive timing, direction, balance, and emotional intent.

Yet many listeners have experienced a puzzling phenomenon: even excellent recordings played on expensive systems can sound disjointed, unstable, or oddly unconvincing. The structure of the music itself has not changed. The musicians performed it correctly, and the recording captured those relationships faithfully.

What has changed is the reproduction.

When the playback system introduces noise, timing errors, or spatial distortions, the relationships embedded in the recording begin to break down. Instruments drift, transients smear, and musical lines lose their clarity. The structure that the composer and performers created is still present in the recording, but it is no longer reaching the listener intact.

To understand why this happens, we must look beyond the music and examine the mechanisms that interfere with it. The next chapter introduces the **Grid Model**, a conceptual framework that helps explain how small disruptions in timing and coherence can cause the musical structure to collapse.

CHAPTER 3

NOISE, THE GRID MODEL, AND WHY MUSIC FALLS APART

Understanding Noise—and What to Do About It

Even the most perfectly placed speakers cannot overcome the limitations of a noisy system. Before true fidelity can emerge, we must confront the invisible forces that obscure it. The next step in the journey toward musical truth begins not with movement or measurement, but with silence—learning to recognize what stands between us and the music itself.

Consider these questions:

- Why do two audio systems, equipped with the same components, sound so dramatically different?
- Why can a small, relatively inexpensive tweak transform one system while having seemingly no effect on another?
- Why has digital audio, despite decades of advancement and clear technical superiority on paper, only recently begun to compete with analog reproduction?
- And finally, where does music live?

The answer to all these questions lies in understanding something the audio industry rarely discusses systematically: the overwhelming impact of noise on music reproduction. You see, music emerges as much from the absence of noise as it does from the presence of sound.

Before exploring speaker placement, it's vital to understand the concept of noise and its impact on sound quality. Reducing noise is at the core of audio system optimization, but how can we eliminate something if we are unaware of its presence? Developing this skill is essential. A good starting point is listening to Mulatu Astatke's "*Tezeta (Nostalgia)*," which serves as an exaggerated example of noise's detrimental effects on music. The track's significant noise levels interfere with the performance, making it difficult to fully appreciate the melodies. Imagine the clarity and beauty of the music without that noise—each note would be more coherent, and the listening experience more immersive. This example underscores the importance of noise reduction. By eliminating each layer of noise, we reveal new levels of clarity and musical detail.

For our purposes, we will define noise as anything that causes the sound to deviate from its intended characteristics regarding content, location, or timing. Consider the following examples:

- **Content Deviation:** Not hearing the full decay of a piano note indicates that the sound is not what it should be, caused by noise affecting note structure.
- **Spatial Deviation:** A singer's voice appearing left instead of the center is an example of sound not originating from where it should be, caused by noise affecting spatial accuracy.
- **Temporal Deviation:** A mistimed treble note represents sound occurring when it shouldn't, resulting from noise disrupting the timing.

Many people believe their audio system doesn't have a noise problem because they don't hear hum, buzz, or hiss from the speakers when music isn't playing. Even placing an ear close to a speaker driver with the volume knob turned up only reveals silence. This is certainly one form of noise that should be addressed if present; a more insidious type of noise is generated when components transmit signals. Often, we are unaware of its presence until it is removed. Consider two examples:

- **Analog Domain — Rumble:** This occurs when the turntable picks up low-frequency content from the music playing through the speakers and feeds it back into the system. We don't hear this "rumble" unless music is playing.
- **Digital Domain — Jitter:** The clock crystal in a DAC is highly sensitive to vibrations, causing jitter. It's well established that jitter adversely affects sound quality, but its impact isn't noticeable until music is played.

In both scenarios, the system harbors noise sources evident only during playback. Various components can introduce different types of noise, and if we want to reveal the music's expressiveness, then it is necessary that we work to eliminate as much noise as possible. Why? Because this noise is removing the "silence" from music.

This brings us to a timeless insight that captures why noise is so destructive to musical expression. Wolfgang Amadeus Mozart is often credited with saying, "The music is not in the notes, but in the silence between them."

He emphasized the importance of the pauses and spaces in music, suggesting that these silences contribute as much to the musical experience as the notes themselves. To understand how noise interferes with musical expression, we must first examine the structure of a musical note.

When a guitar string is plucked or a piano key struck, the resulting sound produces a beautiful tone comprising a fundamental frequency accompanied by a series of harmonics. While discussions about music often focus on aspects such as frequency response and tonality, the temporal characteristics of a note are equally important.

A musical note can be broken down into four distinct stages:

- **Attack:** The initial onset of the note, occurring rapidly as the instrument produces sound.
- **Decay:** The subsequent decrease in amplitude following the attack, leading to the sustain level.
- **Sustain:** A period during which the note is held at a consistent amplitude.
- **Release:** The final phase where the sound diminishes into silence after the note is released.

These stages are collectively known as the ADSR envelope, a fundamental concept in sound design and music production. The "silence between notes" refers to the release phase and the subsequent quiet before the following note begins. This silence plays an indispensable role in defining rhythm, phrasing, and the overall expressiveness of a piece of music. By thoughtfully incorporating these pauses, composers and musicians can create tension, anticipation, and resolution, enriching the listener's experience. Recognizing the significance of the spaces between notes allows for a deeper appreciation of the nuances and emotional impact inherent in music.

The illustration on the following page shows the ADSR envelope (note anatomy) for a single note followed by the start of a second note. The X-axis denotes time and the Y-axis denoted amplitude (loudness). Each of the five phases are labeled for clarity. Observe that the attack occurs very quickly but does take a small amount of time to reach its maximum loudness. Regarding speaker placement, I would refer to this as the system's transient response. How quickly do we perceive the attack of the note? During the decay and sustain phase, we will determine the loudness and pitch of the note. The release phase starts when the musician releases their finger from the key or stops blowing into the instrument. This is where the note will decay to silence or the system's noise floor. And finally, after the note is fully decayed, we have the silence before the following note.

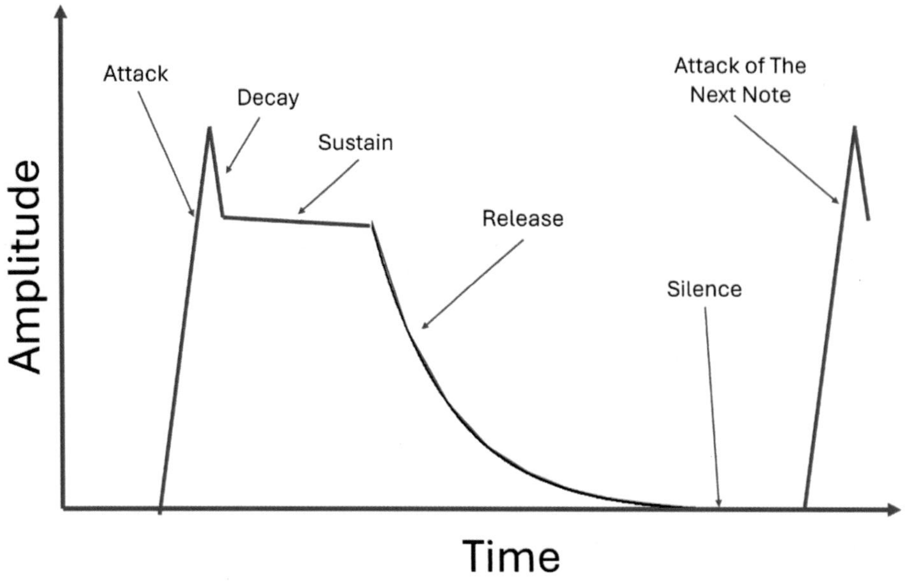

These temporal stages become even more meaningful when we consider the articulation choices composers specify. The pioneers of music notation developed various symbols to indicate how notes should be executed—particularly regarding their attack and the duration of silence between them. These symbols, known as articulation markings, guide musicians in interpreting the composer's intentions. Common articulations include:

- **Legato:** Notes are played smoothly and connected, with no discernible break between them.
- **Fermata:** A note or rest is held longer than its standard duration, at the performer's discretion.
- **Tenuto:** Indicates that a note should be sustained for its full value, sometimes slightly longer, often with a gentle emphasis.
- **Spiccato:** A bowing technique in string playing where notes are played in a detached, bouncing manner.
- **Staccato:** Notes are performed sharply detached or separated, creating a crisp, distinct sound.

The images on the following page were captured playing three successive notes on a piano.

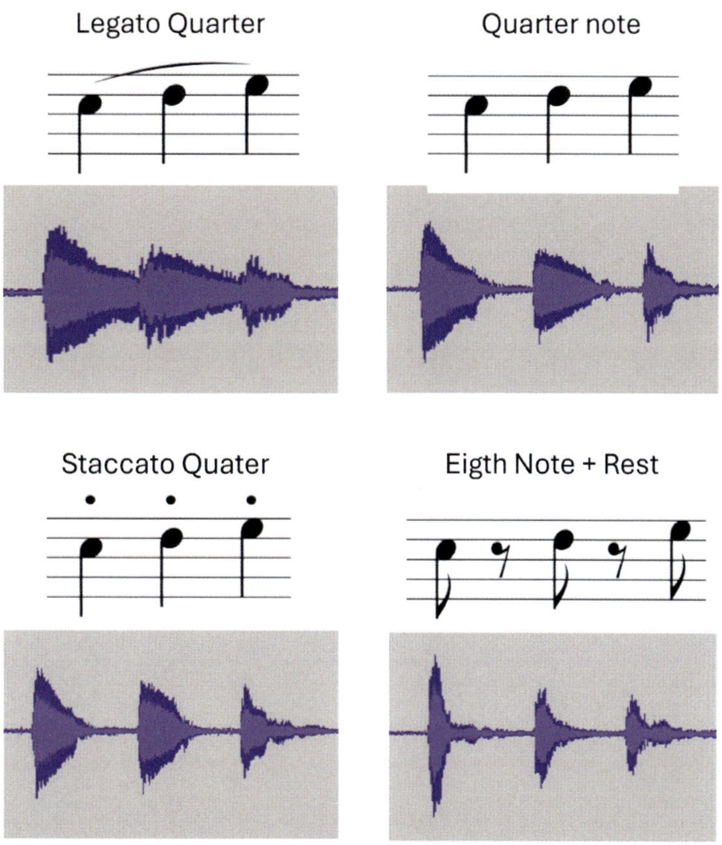

Each of the four images represents a different note attack and spacing before the next note. In the image showing legato notice how the ADSR envelope (shown below the notes) of each note blends into the next note. There is no silence between them. Compare this to the envelope of the staccato notes where we have a substantial amount of silence prior to the following note.

High-resolution music playback systems should accurately reproduce these articulations, allowing listeners to discern the subtle differences the composer intended. Noise within the playback system can obscure these nuances, leading to a misrepresentation of the music. For instance, in Ludwig van Beethoven's *Piano Concerto No. 1 in C major, Op. 15*, there is a passage at 3:42 where the pianist plays a series of descending arpeggios. On many audio systems, this sequence is rendered as if marked *legato*, with notes smoothly connected. This effect results from system noise masking the intended articulations. Such systems are often described as "musical," but a more accurate term might be "lyrical." It is essential to recognize that not all musical passages are meant to be lyrical; some require distinct articulations to convey the composer's intent. Reducing noise in playback systems is essential to reproducing the music as it was written and performed.

To see how noise alters these articulations in practice, let's examine a specific example of musical notes, focusing on their attack, release, and the following silence. Consider the opening six notes of *Rossini's L'Italiana in Algeri* overture. In a high-quality recording—such as the one by the Orpheus Chamber Orchestra—each note is distinct, with clear attacks, full sustains, natural release, and noticeable silences between them. This clarity allows listeners to appreciate the precise articulation the composer intended.

In contrast, a recording plagued by higher noise levels will obscure these details. The increased noise can mask the release of one note and the attack of the next, resulting in a muddled sound where notes overlap unnaturally. This results in diminished clarity and a loss of the music's intended expressiveness. By comparing these two scenarios, we can see that noise significantly impacts the quality of note structure and therefore the music's integrity. **Figure 1** illustrates the ADSR envelope of the first six notes in *L'Italiana in Algeri* and how noise affects it. In this figure, the Y-axis represents loudness, and the X-axis represents time. Observe the differences in note structure. The Orpheus recording exhibits a much more precise transient attack and release and shows significantly more silence before the following attack. Counting the notes, we see only five in the alternate recording—one note has been lost in the noise.

Figure 1: ADSR Envelope of First Six Notes of *L'Italiana in Algeri*

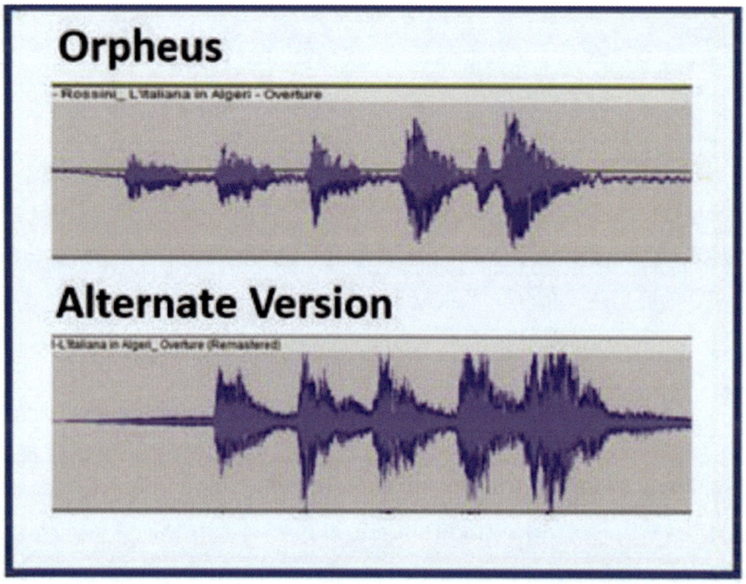

While we cannot control how recordings are made, these graphs help us visualize and understand the concept of noise. They illustrate how noise can degrade an impulse's quick, clearly defined rise, its distinct decay profile, and the clean separation between notes. Imagine both graphs are from the same recording of *L'Italiana in Algeri,* captured by recording our stereo system playing this piece in our living room. The top graph represents an optimized system with reduced noise, while the bottom graph depicts the unoptimized system. This visual comparison demonstrates the effect of reducing noise in our system and underscores its importance in maintaining the music's integrity. Transitioning the sound from the bottom graph to the top graph is precisely what happens when following the processes outlined in this guide. To better understand why these musical effects occur, we can turn to a visual model that captures the mechanics behind them.

Grid Lines: A Model for Visualizing Noise

The grid model provides a straightforward method to help visualize the effects of slight speaker adjustments or reductions in a system's noise floor. To state the model succinctly, each line of the grid represents a position where one of the speaker's six degrees of freedom (e.g., front-to-back, lateral, height, toe-in, rake, and azimuth) is correctly aligned. Achieving alignment along these grid lines corresponds to reaching a local minimum of acoustic noise. The thickness of the lines and the opacity/transparency between them symbolize the presence of noise.

What is a Model, and Why Do We Need One?

A model is an intellectual construct used to help visualize and explain observed phenomena. Scientists routinely employ models to conceptualize complex ideas. For instance, the Bohr model of the atom, introduced in 1913 by Danish physicist Niels Bohr, depicts electrons orbiting the nucleus in defined paths, like planets around the sun. While this model has limitations, it provided a foundational framework for understanding atomic structure and facilitated future advancements in quantum mechanics.

Similarly, in audio system optimization, a model is useful as it can help us visualize and understand system behavior. Stated simply, the model represents sound emanating from speakers as a series of grids.

The grid model emerged from observations during speaker system setups. It's theoretical framework helps visualize how sound behaves when aligning speakers or tuning system components. Notably, explaining why small adjustments in speaker positioning can result in significant improvements in sound clarity. As speakers are moved incrementally in a particular direction, the audio will become progressively clearer until it reaches an optimal point—after which further movement will cause the sound to lose focus. This pattern of clarity and obfuscation resembles a grid, where specific positions yield optimal sound quality.

Let's start by considering the most simple application: speaker alignment within a room. In the representation pictured on the following page, the blue lines represent the room's acoustic properties, while the red lines represent the speakers' output characteristics. The goal is to align these grids in a way that minimizes acoustic noise. As shown in this example, if the grids are misaligned in front-to-back and lateral positions, it indicates excess acoustic noise. By making small adjustments—such as moving the speaker (red grid) backward to align the front-to-back position and shifting it laterally to the left to align the lateral position—we can achieve better alignment, resulting in reduced noise and improved sound clarity. Alternatively, moving the speaker forward and to the right could also achieve alignment, though it might require more significant adjustments.

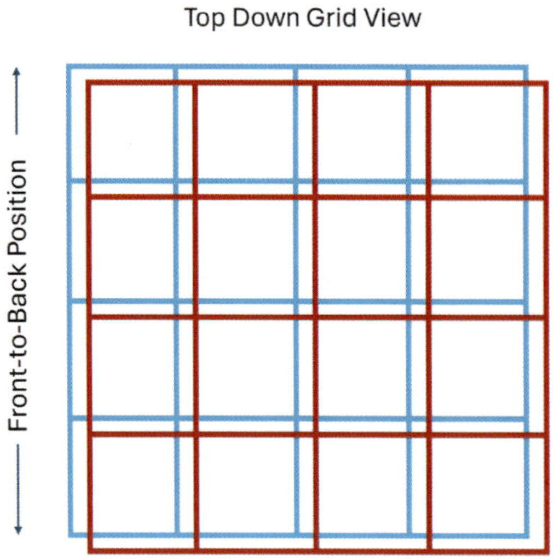

The grid concept is likely new and unfamiliar, so we need to explore it further by examining some examples. Consider two speakers. Visualize each speaker's sound forming a grid. Every speaker and each of its drivers has a grid, with higher frequencies producing smaller grids and lower frequencies producing larger ones. The speaker setup section will discuss this in more detail, but for now, refer to the grid lines shown in Figure 2. The illustrations show the grids (sound) emanating from the left and right speakers and intersecting at the listener's ears. In the left image, the grids are not aligned, while in the right image, they are aligned.

The leading edge of the sound impulse (attack) will be smeared in time if the grids are not aligned. This misalignment also results in lost information—content that exists but is obscured and hidden between the grid lines. Misalignment further causes truncated decay and a loss of silence between notes. This grid visualization model allows us to see the same three effects of noise that we identified when examining the ADSR envelope: poorly defined transients, unnatural release, and a lack of silence between notes.

Figure 2: Grid Lines

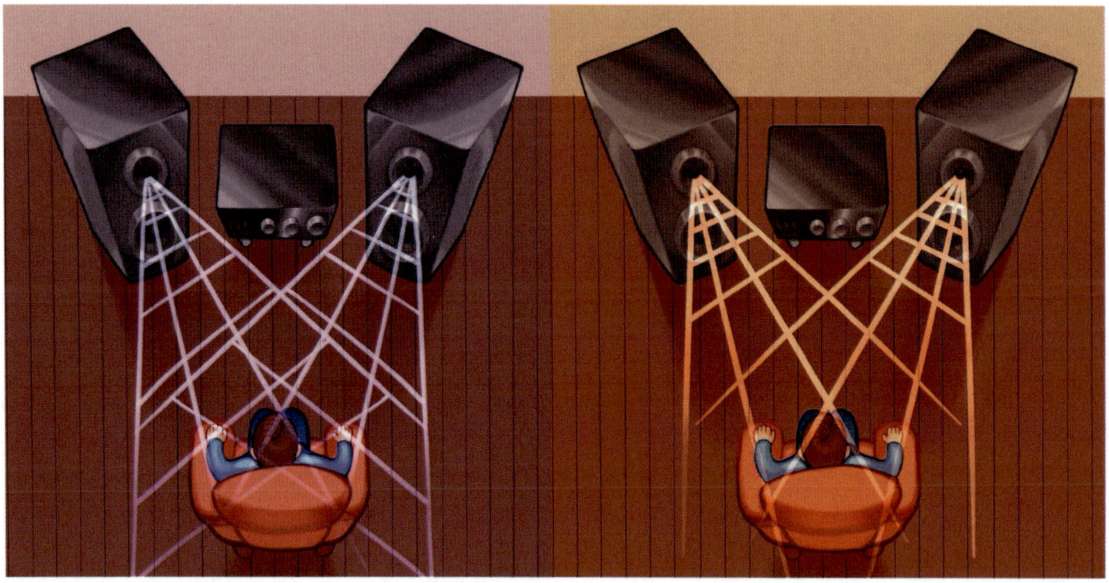

Additionally, the room has its own grid lines, resulting from sound waves emitted by the speakers reflecting off the walls and returning to our ears. Aligning the room's grid with the speakers' grids is essential, as misalignment can cause the music to sound congested, unnaturally resonant, edgy, fatiguing, or overly bright. Reducing acoustic noise to the lowest possible level is necessary to eliminate these maladies, and the speaker setup portion of this guide focuses on achieving that goal. It's also important to recognize that the room's grid lines are not fixed; they can change with modifications to the system, such as replacing speakers. Therefore, new speakers cannot simply be placed in the same position as the old ones and achieve optimal sound quality, as many are already aware. Obviously, this applies to replacing any component within the system.

Once the speaker and room grids are aligned, we will perceive very low acoustic noise. The removal of time smearing enables dynamic contrast and delicacy. Acoustic guitar plucks become crisp, and fine detail emerges. Unfortunately, the prevailing belief in home music reproduction is that acoustic panels are necessary to cure acoustic noise and achieve great sound reproduction in a small room. This is not true. Excellent sound can be achieved in most rooms with no special treatments. While a few panels may offer some improvement, the results after proper grid alignment are far more impressive.

Grid lines have a finite width that varies according to the amount of mechanical noise in the system. As illustrated above, greater mechanical noise produces wider, fuzzier grid lines, resulting in two audible effects. First, a wider grid line will obscure the leading-edge transient, since we cannot precisely determine where two grids are meant to intersect. A wide grid line introduces uncertainty in the speaker's position, inhibiting our ability to perceive the music's precise timing cues. The result is a loss of dynamic shading, making the music sound slightly dull or uninspired. Second, a wide grid line causes notes (tones) to sound less pure and, when taken to the extreme, results in a harsh, raspy character.

The space between the grid lines can range from translucent to opaque, which affects our ability to clearly "see" the music. If a semi-transparent haze fills these spaces, instruments will exhibit less realistic decay and reduced quietness between notes. I associate this with electrical noise. This "grid haze" compresses dynamics, as the noise prevents notes from decaying fully. Consequently, the distance between a musical *pianississimo (ppp)* and *fortississimo (fff)* is reduced. The wider grid lines and shading between grids limit the potential for sound quality improvement beyond a certain level of accuracy.

The drawing below depicts grid line width and shading. As the grid lines become narrower, it becomes easier to identify the exact crossing point. The more precisely we can see—or hear—this intersection, the more accurately we can align the system's timing. This leads naturally to the concept of transparency, which the grid model illustrates vividly.

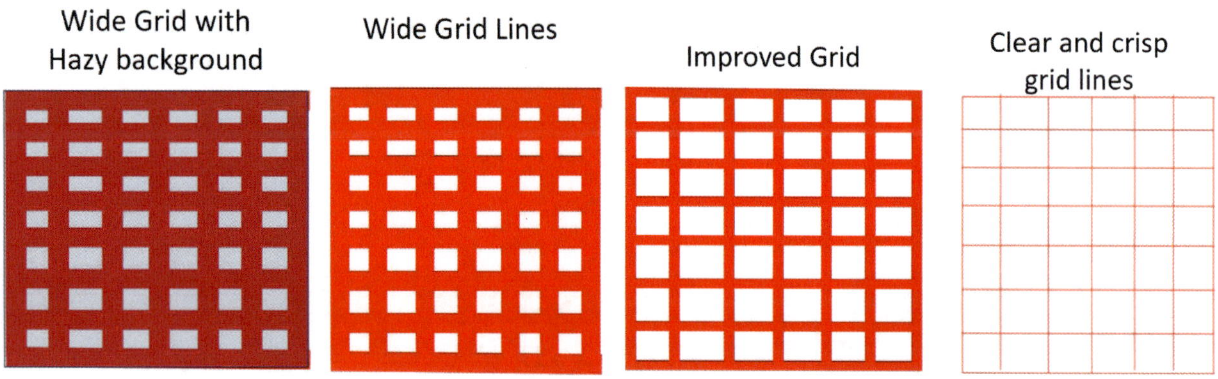

Using the grid model, it becomes clear (pun intended) what is meant by the term *transparency*. The image below illustrates this concept: in the picture on the left, the grid lines are thick, and the haze between them almost completely obscures the image. In contrast, the picture on the right clearly reveals the portrait of Beethoven.

The width of the grid lines determines the positional accuracy we can perceive, which in turn dictates how precisely we can align the speakers. Therefore, the amount of noise in the system limits the absolute sound quality that can be achieved. The example below shows how sound quality becomes truncated when we cannot listen deeply enough into the system to perceive finer positional accuracy. With the model established, we can now explore how real systems behave as noise is reduced.

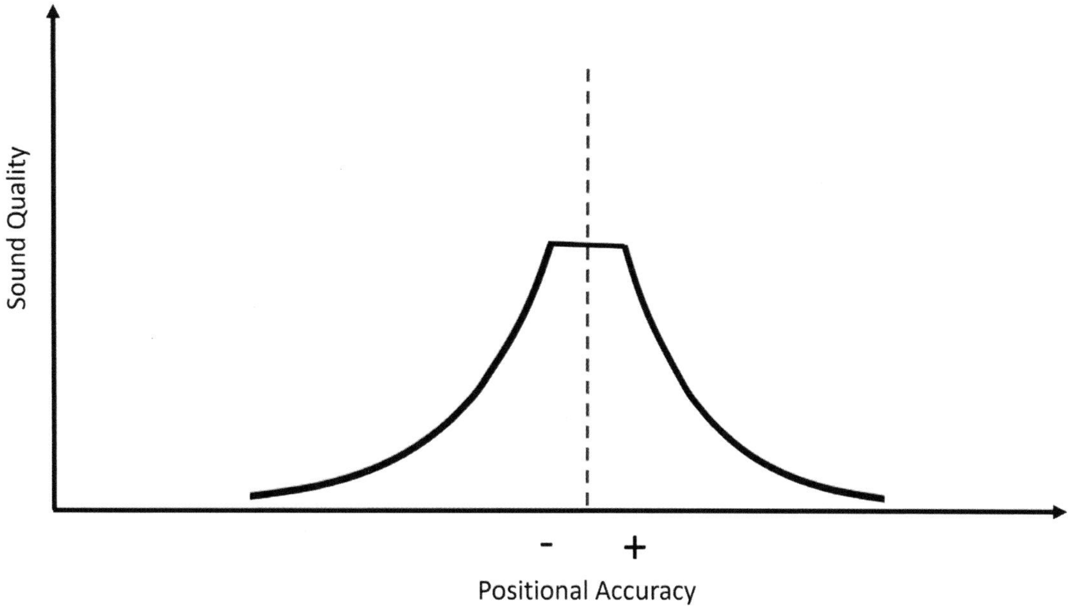

Removing Noise and Exposing Misaligned Grids

If there is noise in the system, the space between the grids will range from opaque to translucent. Although the goal is true transparency, it is not fully achievable—we can only continue removing layers of noise. Once a system is dialed in, the grids will sound aligned; however, if something is added to the system that further reduces noise, the grid lines will become narrower. Grids that were once thought to be aligned will now reveal misalignment. This newly illuminated discrepancy can make the system sound bright, harsh, or murky—artifacts that were previously smoothed over. We were simply unaware of the problem.

Does this mean that our change is bad? No, not at all. The modification has lowered the noise floor and increased the system's musical resolution. We simply need to realign to the newly exposed grid lines, which will eliminate the brightness, harshness, or murkiness—yielding a higher degree of true musicality in tone, timing, and dynamics.

The depiction below shows one grid with very wide lines and another where the spaces between the lines are fairly opaque. Removing either type of noise reveals the third box, where the previously obscured grid lines become visible. We then need to align the grids to resolve any acoustic anomalies, such as brightness.

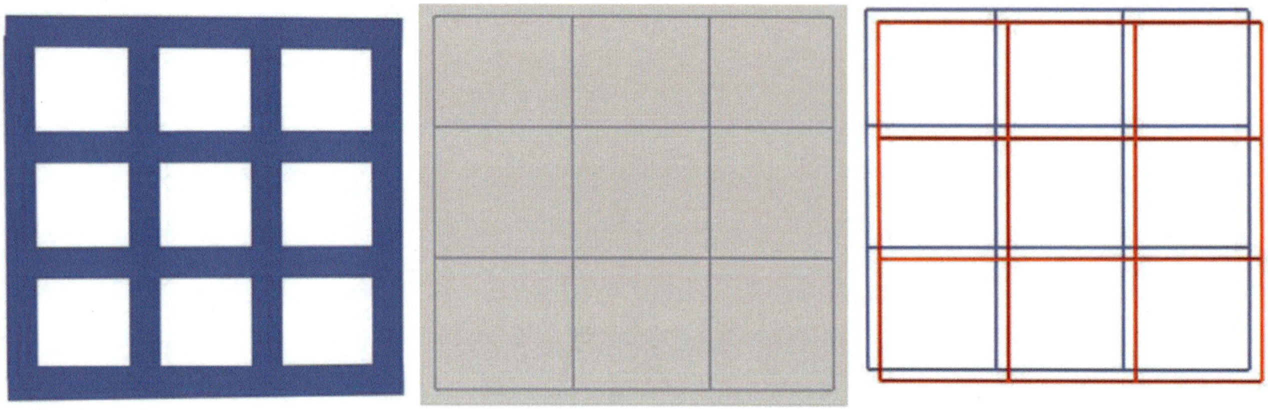

The Grid Model in Practice

I mentioned earlier that I experimented with two different preamps during my third session with Stirling. Utilizing the speaker setup process detailed in this guide, we initially set up the system with a very good solid-state preamp. The piano in *Beethoven's Piano Concerto No. 1* sounded excellent, with even note spacing and articulation. The supporting orchestra was organized and dynamic. Overall, the sound quality was high, and most listeners would have been thrilled with it—as I was.

We then removed the first preamp and inserted a different, much higher-quality model. One would expect the second preamp to be a slam dunk, with everything sounding noticeably better. This was not the case. Piano note articulation was indeed improved, and the clarity of notes from both the piano and orchestra was strikingly better. The second preamp had a much lower noise floor, resulting in remarkable transparency. Sadly, the music itself fell apart. It lacked drive, sounded splashy, and the musicians seemed out of sync—as if a high school band had replaced the professional orchestra.

Applying the grid model helps explain what happened with the supposedly superior preamp. The new preamp removed haze and produced a much finer grid line, revealing that the system's grids were misaligned and exposing timing errors that had previously been masked. The result was musical disorganization. Fortunately, only minor adjustments were needed to restore coherence—several small bumps outward and forward, a subtle rake change, and a tiny toe-in nudge. The speaker appeared to be in the same position, but it was not in the same position acoustically. After these small movements, the new preamp's true capability emerged, and the musical performance was elevated several notches.

This experience illustrates a core truth of high-fidelity audio: lowering the noise floor doesn't just reveal more detail—it exposes every underlying misalignment in the system. As the grid becomes finer, even tiny errors become audible, and the system must be re-balanced to match its new level of resolution. With that in mind, we can now turn to the practical question: where does this noise come from, and how do we remove it?

Steps to Reduce Noise in the System

We have defined noise, outlined the components of a musical note, visualized how noise degrades note structure, and built a model to visualize noise in our system. To understand where this noise originates, let's break down our system into smaller parts. A home music system comprises two primary subsystems:

- **Electrical System:** This includes the incoming power supply and electronic components such as CD players, DACs, turntables, pre-amplifiers, and amplifiers.
- **Acoustic System:** This consists of the speakers and the room in which they are situated.

Both subsystems can introduce unwanted noise, diminishing the musical experience. To effectively address these issues, it is helpful to categorize the noise into three types:

- **Acoustic Noise:** Unwanted sounds from the environment, such as HVAC systems or external traffic, and acoustic clashing from misaligned speakers.
- **Electrical Noise:** Interference within the electronic components, often resulting from grounding issues or electromagnetic interference.
- **Mechanical Noise:** Vibrations or sounds produced by mechanical components affecting sensitive signals.

Understanding these categories allows us to systematically identify and mitigate the specific types of noise affecting the audio systems. This approach is paramount for preserving the integrity of note structures and ensuring a high-fidelity listening experience. We will consider each one in turn and discover what can be done to reduce noise.

Acoustic Noise

Acoustic noise can originate from outside the system or be generated within it. We should seek to silence both sources. Examples of external noise include air conditioning, refrigerators, traffic, upstairs neighbors, and family activities. There is only so much we can do to silence this kind of noise unless we are willing to invest in a dedicated listening room. Although, we can do a great deal to reduce the acoustic noise generated within our system—and that will be our focus.

When you think of acoustic noise from within the system, you might imagine things like transformer hum. However, this is actually mechanical noise, not what I mean by acoustic noise. Acoustic noise comes from speaker misalignment. This type of noise creates cognitive dissonance, pointing us to realize that we are listening to a high-fidelity system rather than real performers. Real people, in real acoustic spaces, do not generate this kind of noise. So, what is it?

Recall the section in Chapter One titled "Stereophonic Sound and ITD/ILD from Two Sound Sources," where we examined the graphic below. It shows the sound from the left and right speakers reaching each ear at different times. When a musical event occurs, each ear hears two clicks instead of just one. This is a clear example of acoustic noise. When the speakers are perfectly aligned this is the lowest acoustic noise. Any misalignment will produce extraneous noise. This will become clearer by looking at an example.

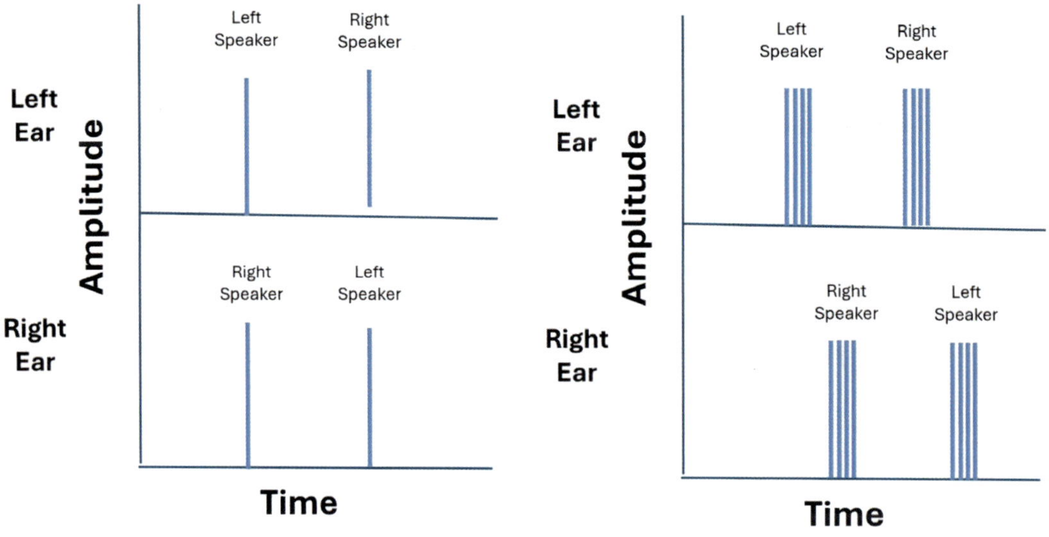

Consider the effect of a simple three-way speaker. Each speaker has a tweeter, a midrange, and a woofer. Suppose we have not utilized the processes outlined in this guide and instead place the speakers in a visually appealing position. In this case, the tweeter and midrange drivers are misaligned, the midrange and woofer are misaligned, and the left and right speakers are misaligned. Each pair of drivers produces an extra click: the left speaker will have four clicks instead of one, the right speaker will have four clicks instead of one, and the pair together will have three clicks instead of one. Not counting the room's effect, each ear will experience eleven clicks instead of just two—many times more than it should! This is acoustic noise as shown on the right side of the illustration above.

With this understanding, the grid model begins to make sense. Each time we snap a speaker into alignment along a grid line, we position it so that we hear one click instead of two. Each successful alignment reduces the acoustic noise within the system. This is the essence of the techniques this guide elucidates.

Electrical Noise

The electrical system feeding your gear is fundamental: the power from the wall is the raw material electronics use to move a speaker's drivers and produce sound. Unfortunately, many sources of **electrical noise** can degrade performance. Noise can enter through the AC line's *hot* leg, the ground line, or from the gear's own power supplies. Additional interference comes from EMI/RFI generated by Wi-Fi, cell towers, and other household devices. All of it eventually contaminates the low-level signal traveling through cables and electronics. Below are the most common sources—and steps you can take to address them:

- **Noise on the Hot Leg (AC Line):** Reducing incoming AC noise is valuable but requires care. Many power conditioners claim to clean the line while also providing extra outlets. They can improve clarity and detail but sometimes add mechanical noise and suppress transient response. Always listen critically—consider both the positives and any new negatives introduced after plugging electronic components into a power conditioner.

- **Ground Noise:** The ground is the "zero point" against which signals are referenced, and improvements here can yield big results. Start by ensuring all ground connections are secure and of equal path length. For more thorough work, an electrician can tighten breaker box connections and inspect the house ground rod. Passive "grounding boxes" are also highly effective at lowering the noise floor and increasing perceived quietness between notes.

- **EMI/RFI Interference:** This is more complex. Cables act as antennas, so shielding is critical. Use shielded cables with the shield tied to ground only at the sending end; grounding both ends creates a conduit for noise. Some accessories absorb high-frequency interference when placed near cables or components. Evaluate these in your own system—don't rely on marketing claims.

- **Ethernet Noise:** Ethernet connections are a major source of contamination. Avoid using them unless necessary. If streaming is required, mitigate noise with switches, shielded Ethernet cables, and filters. Achieving acceptable performance often requires multiple layers of filtering.

- **Dirty Electrical Contacts:** Every cable connection transfers signal across contact points (RCA, XLR, IEC). Treating these with a product like Super TMD or DeoxIT reduces signal loss and noise. Treat both interconnects and IEC power connections.

- **Outlet Integrity:** Verify outlets are wired correctly, with proper polarity and ground. A simple tester can confirm this for minimal cost.

- **Dedicated AC Lines:** Dedicated circuits isolate your audio system from household appliances on the same circuit, lowering noise and reducing circuit impedance. Ideally, use separate lines for front-end gear and amplifiers. 10-gauge wire is preferable to 12-gauge: it offers lower impedance and greater current capacity, with little added cost or labor. To further reduce noise run the wire through steel conduit.

Dedicated lines do many good things, but there is a pitfall—multiple dedicated lines must be installed carefully. Ensure each run back to the breaker box uses the same gauge and equal length, or you risk a **ground loop**, which will introduce a destructive 60/120Hz hum into your system.

Mechanical Noise

The system's mechanical noise also needs to be addressed. Another word for mechanical noise is *vibration*. Vibration is bad. It smears the "grid lines" and imbues the sound with a "bloom," making the tone less pure. This effect can be nonlinear with frequency—sometimes affecting the high frequencies and, in other cases, the mid or low frequencies. We want to prevent mechanical energy from entering the system and quickly remove any vibrational energy generated within it. The goal is mechanical stability.

- **Leveling:** The rack or stand on which all electronics rest must be level, and each shelf should also be level. Every component on each shelf should be level as well. We do not want anything to "wobble," even slightly. Ideally, everything should rest on adjustable feet to achieve a perfectly level position. If adjustable feet are unavailable, you can use shims (e.g., paper or wood) under each foot until the component is level.

- **Draining Vibrational Energy:** Each component needs a mechanism to drain vibrational energy. Most components come with soft, rubber-like feet that isolate the component from the shelf but prevent mechanical energy drainage. Footers come in many forms and follow different design philosophies. The footer must contact the chassis, transfer vibrational energy, and dissipate it. Something as simple as a wooden block can be effective.

- **Avoid Stacking Components:** Components should not be stacked on one another. Stacking couples them together, allowing vibrational energy from one component to affect another—resulting in degraded performance.

- **Tighten Screws:** Loose screws cause unwanted vibration. All screws in the casework should be tight. Check the screws that secure the drivers to the speaker cabinet as well. Do not overtighten them but ensure they all have equal torque. You might be surprised to find some slightly loose screws if your speakers are a few years old.

- **Power and Cabling:** A major source of vibrational energy is the AC power. All power cords should be secure. If they sag, support them with a wood block or something similar. Some damping of power cords (and cables in general) can be beneficial. Avoid laying interconnects on top of power cords, regardless of orientation.

This may all seem obsessive. Not long ago, I felt the same way. It is rarely a single adjustment that makes or breaks the system, but the cumulative effect of many small improvements is dramatic and transformative. Once everything is addressed, the difference is shocking when the music starts to play. The sound becomes more transparent, dynamic, and alive. The good news is that most of these items require more time than money—many are free or very inexpensive, especially compared to the cost of even a single component in a system. Cleaning up all mechanical noise will make timing inaccuracies easier to hear, making it easier to achieve perfect speaker alignment.

Assessing Noise in the System

Two questions are looming: *"How do I know how much noise my system has? And how do I know when it is good?"* To answer these, we will use four reference tracks: "Lucky," "The White Wind," *L'Italiana in Algeri*, and *Schubert's Piano Sonata No. 20 in A Major D. 959.* Each track has a special trait that reveals something different about noise. Let's consider each one.

1. "Lucky" — Revealing Mechanical Resonance & Nonlinear Distortion

"Lucky" is an excellent tool for exposing mechanical resonance. At the beginning of "Lucky," Kat is "Hoooo'ing." She is dubbed over herself and singing two notes—G♯4 (415 Hz) and C5 (523 Hz). By loading "Lucky" into Audacity, splitting the channels, and examining the frequency spectrum of the first second of music, we can observe two fundamental frequency peaks and their corresponding harmonic spectra. Interestingly, the 415 Hz peak is stronger in the left channel, while the 523 Hz peak is more prominent in the right channel. This can be confirmed by carefully listening to the track with headphones.

"Lucky" Right Channel

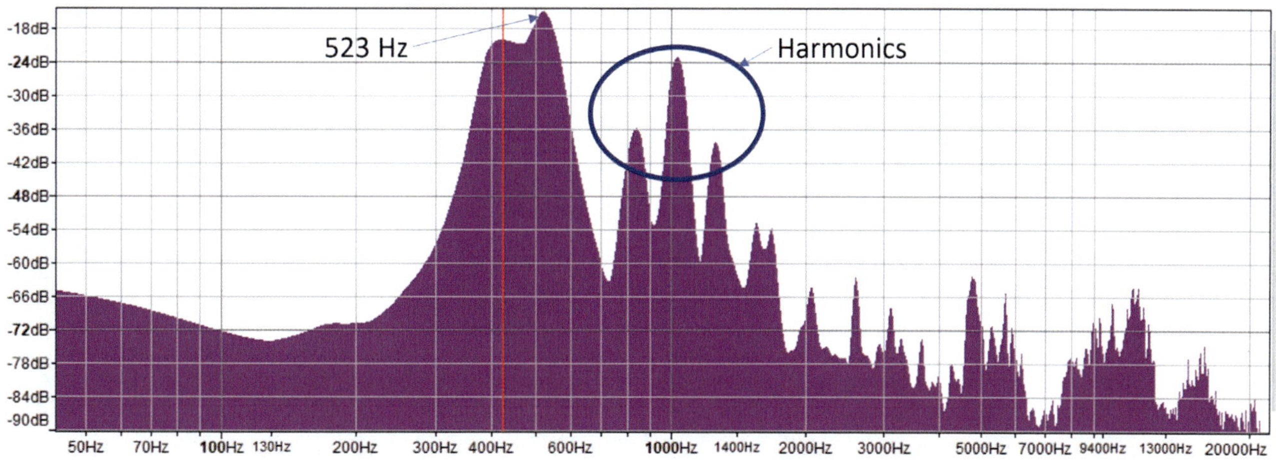

"Lucky" Left Channel

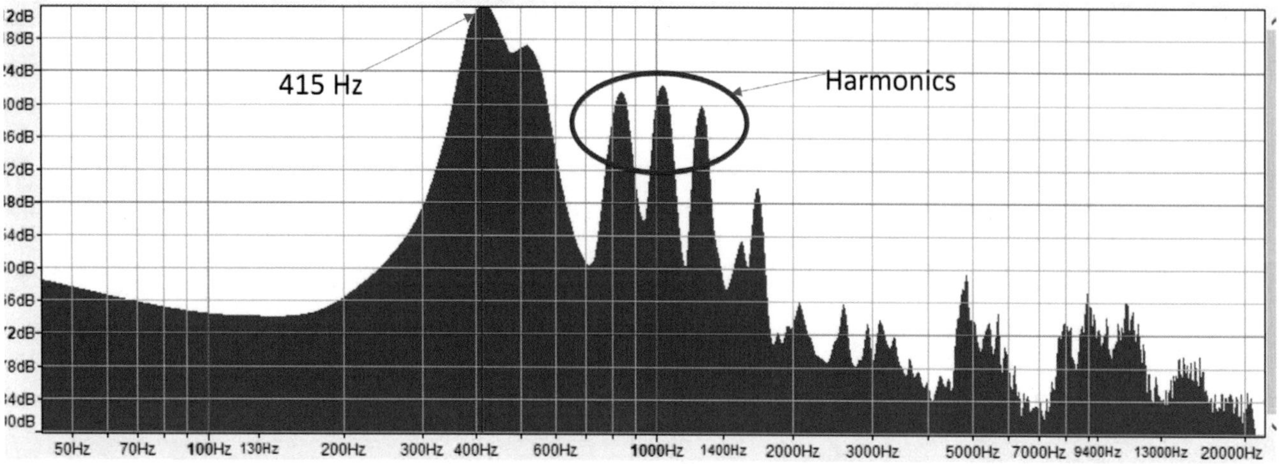

While this is an interesting feature and observation about "Lucky," you may wonder how it helps reveal anything about the system's noise floor. Play and listen to the reproduced sound in "Lucky's" opening seconds and assess what you hear. The more noise in the system, the bigger, blurrier, and more diffuse the *hooo'ing* sounds will be. The tone may also take on a raspy character. In a noisy system, it can be very unpleasant to listen to, and you will likely want to lower the volume.

As mechanical noise is removed from the system, the *hooo'ing* becomes increasingly localized to a central image. It also becomes more tonally pure and less raspy. A point is eventually reached where enough noise has been removed that—almost as if by magic—the large central image splits into two distinct images. You will now hear one singer slightly to the left of center and another slightly to the right. Continue reducing noise, and Kat's *hooo'ing* becomes beautifully focused and deeply pleasant to listen to.

***NOTE**: The "hooo'ing" in Lucky exposes non-linearities in the system. Distortion! There are two primary types of distortion in audio: Harmonic and Intermodulation. Harmonic distortion is generally regarded as benign, and some people like additional harmonic distortion as it adds a "fullness" or "lushness" to the sound. Intermodulation distortion, on the other hand, is universally bad. The two frequencies sung at 413Hz and 523Hz expose the intermodulation distortion within our system. This is what causes the large diffuse sound field and the raspy tone. No amount of precision in speaker placement can fix this.*

Illustration of Flowing Wind

2. "The White Winds / The White Boat" — Revealing Electrical Noise

Andreas Vollenweider's "The White Winds/The White Boat," from the album *White Winds,* is an excellent tool for identifying noise in your audio system. This track highlights the deleterious effects that noise has on sound and illuminates the path toward noise reduction. Observations from many different systems reveal that removing noise follows a specific progression.

In the beginning, music gains body—most notably in vocal tracks. Next, the sound becomes less grainy and much smoother. Then, at a certain point, improved dynamics become easy to notice: the difference between quiet sounds and slightly louder ones becomes more distinct. This is often the first stage where emotion is truly unlocked from a vocal performance. The final stage is a gain in spatial resolution, where the sound becomes more three-dimensional. As we continue lowering the noise floor, we gain additional body, smoothness, dynamics, and dimensionality in the sound. This is demonstrated in the graph on the following page. Notice that in a noisy system, initial reductions in noise produce relatively small improvements in sound quality. However, as we achieve further reductions in the noise floor, the perceived improvements become increasingly significant.

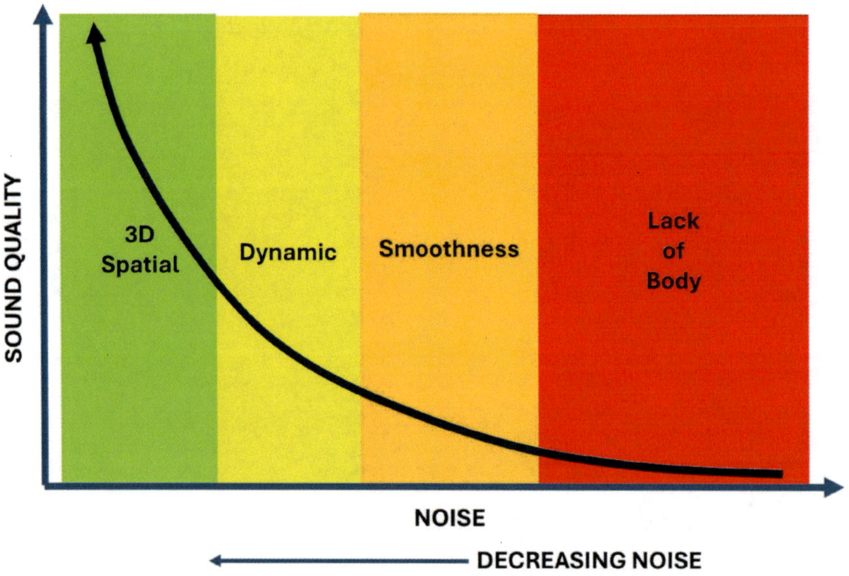

"The White Winds/The White Boat" begins with the sound of—appropriately—wind, which makes it ideal for this purpose. Because wind is a natural sound, it allows us to focus on detecting unwanted noise without the distraction of melodies or lyrics. By carefully listening to the track's introduction, you can assess the clarity and purity of your system's sound, making it easier to identify the effects of electrical noise.

When listening to the wind in "The White Winds/The White Boat," I like to imagine standing on a cold mountain under a gray sky, looking through a great passage carved between high stone walls—something straight out of J.R.R. Tolkien's *The Lord of the Rings*. As I explore the passage, the wind howls and swirls around me. Some of the traits you should listen for in this track are:

- • The wind should have **body.**
- • The wind should sound **smooth and flowing.**
- • The wind should be **dynamic**—rising and falling in intensity.
- • The wind should sound **three-dimensional,** with a sense of depth.

When a system contains a significant amount of electrical noise, the wind will sound whistly, grainy, dynamically compressed, and two-dimensional. We will consider each of these characteristics in turn.

Noise robs the sound of its body. Sound has *body* when the fundamental and lower harmonic frequencies are strong and well defined. Electrical noise adds energy to the upper harmonics, giving the sound a slightly higher perceived pitch. If the system has excessive noise, the wind will take on a "whistling" quality.

To understand this effect, use your lungs and mouth to imitate the sound of wind. First, keep your lips relatively close together, take a deep breath, and blow. Use your tongue to modulate the frequency and notice how high-pitched and "whistly" the air sounds. Next, repeat the process, but this time keep your mouth open. Again, use your facial muscles and tongue to modulate the frequency. Notice that the sound is fuller and deeper. This simple experiment illustrates the effect of removing noise in "The White Winds/The White Boat."

Noise also makes the sound grainy, glitchy, and splotchy across the soundstage. As the wind swirls from right to left and back again, breaks or gaps may appear in the motion of the sound. Noise gives the wind a rough texture; it doesn't flow or swirl like real wind moving through a mountain pass. It lacks the natural smoothness that true wind possesses. The most obvious place to hear this is at 6 seconds where the wind moves from right to left and changes direction at the left loudspeaker. If noise is present, the wind will seem to reach the left speaker, stop, get stuck, and abruptly return to the right. In a clean system, it should make a smooth, continuous loop before flowing back to the right.

Noise also robs the system of its dynamics. When listening to "The White Winds/The White Boat," we should perceive the wind changing in loudness or volume. Sometimes it is quiet; other times, it is louder. These changes can be gradual or sudden. Listen carefully to how much variation in loudness you perceive—the less noise in the system, the more forceful and dramatic the wind will sound.

Noise makes the wind sound two-dimensional or flat, like a picture of wind stretched between the speakers. Removing noise adds a third dimension to the sound. As the wind swirls from right to left and back again, we should hear it filling a three-dimensional space behind the speakers, rather than resembling a two-dimensional projection. Again, imagine standing before the mountain pass—you would perceive the wind both in front of you and extending into the distance.

Moving past the wind, we encounter the white boat—a sonic image of a woman rowing across the water. We hear the oars gliding across the surface, water splashing against the hull, and droplets falling from the blades. In a noisy system, this presentation will sound one-dimensional, as though all the information has been compressed into a line between the speakers. As noise is reduced, the water becomes more three-dimensional. We can "see" the oar dipping into the depth of the water and rising back above it. We can hear droplets falling from the oar and splashing back into the sea. Pushing the system to even lower noise levels reveals every nuance of the water—it becomes almost tactile.

At first, you may not have a clear reference for *"The White Winds/The White Boat,"* making it difficult to know whether what you're hearing is correct. If the wind sounds whistly, flat, or glitchy, there is certainly unwanted noise in the system. If the water sounds like digital hash, the noise floor remains too high. *"The White Winds/The White Boat"* is a valuable tool for comparing products that claim to reduce noise, such as grounding boxes. Play the track with and without the device, and any genuine improvement in sound quality should be immediately apparent.

3. L'Italiana in Algeri — Revealing Silence, Decay, and Microdynamics

Our third assessment tool is *L'Italiana in Algeri.* The recording by The Orpheus Chamber Orchestra (listed in the reference songs) has an extremely low noise floor. The attack, release, and silence between notes in the first 25 seconds should be obvious. If the system contains excessive noise, the space between notes will be missing. Removing noise reveals the silence; removing even more allows the plucked strings' release to come through before the next note begins. As we continue reducing the noise floor, the strings take on a bounce. You should hear that some phrases are louder while others are softer—and even within a single phrase, certain notes vary in intensity. The lower the noise floor, the more musically expressive the introduction becomes.

4. Schubert Piano Sonata No. 20 in A Major, D. 959 — Revealing Temporal Distortion & Pace

The final assessment tool for system noise is *Schubert's Piano Sonata No. 20.* The "pace" of this piece should feel achingly slow. Notably, noise always makes music sound too fast. The perception of increased tempo likely results from truncated release and a lack of silence between notes. Because of this, one note seems to flow directly into the next, making the tempo appear faster than it truly is. Until you become familiar with this piece, it may be difficult to assess its correct pace. Although, you can use it to evaluate noise differences in your system when comparing products. If you insert a product and the music slows down, that is a good sign. We will revisit the concept of musical pacing when we discuss speaker toe-in.

When a system change is made, these four tracks can provide valuable insight into the device's noise characteristics. Listen to each one carefully and take notes—either on paper or mentally—about the before-and-aftereffects of the change. You may find that a device lowers the electrical noise floor but raises the mechanical noise floor, or vice versa. Ideally, the new component should improve all aspects of sound quality.

Practical Steps to Improve the Electronic System

1. **Polish** the IEC connector pins on all electronic components.
2. **Clean all connections** using 99% isopropyl alcohol and condition them with Super TMD, DeoxIT or another contact enhancer. This includes power connections, interconnects, and speaker terminals. Be cautious—too much product can be detrimental.
3. **Verify AC outlet polarity** is correct. If multiple dedicated lines exist, ensure they are of equal length and connected to the same electrical phase in the breaker box.
4. **Tighten electrical connections** in the service panel. A qualified electrician should perform this.
5. **Inspect the ground rod and connection** to confirm proper contact and address any issues.

6. **Level the equipment rack**, and level each shelf if your rack design allows.
7. **Give each component its own shelf.** Separate power supplies should also have individual shelves.
8. **Use energy-draining feet** (such as small wooden blocks) under each component.
9. **Level each component** precisely on its shelf—the more accurate, the better.
10. **Check for loose screws** on all equipment cases and tighten them if necessary.
11. **Check speaker driver screws** to ensure they are tight and equally torqued.
12. **Damp the top plate** of each component with a small weight— assess if a bag of lead shot or a similar damping strategy reduces the noise floor.
13. **Secure power cords.** Make sure the IEC connectors fit snugly; if they sag, support them with a small wooden block or similar prop. Sleeves designed to tighten IEC connections can also help.
14. **Connect interconnects and speaker cables.** Tighten speaker terminals firmly but avoid overtightening. Keep interconnects as far as possible from power cords. If they must cross, do so at a 90-degree angle. Wooden "bridges" can help maintain separation.
15. **Ensure cables are free from strain and sharp bends.** This may require pulling the rack slightly away from the wall.
16. **Keep the rack low** (under 30 inches) if it sits between the speakers, as it can act as a reflective noise source.
17. **Experiment with products** designed to reduce electrical and mechanical noise. Many are available, and grounding boxes are especially worth testing. Use the music tracks discussed above to determine whether changes are positive or negative.
18. **Address Ethernet noise by installing a low noise switch prior to the streamer, utilizing filters and using a shielded Ethernet cable with the shield attached only at the sending end.**

Noise Floor and Evaluating Changes

Many audiophiles enjoy discovering and experimenting with products that promise enhanced sound quality. Most have also encountered situations where no significant difference was detected after installing the latest device. Every system is unique and functions as its own ecosystem. When evaluating a product that claims to improve performance, consider its position within the overall noise floor. The device may function as intended, but its benefits could be masked by other elements in the system that generate noise. The following three examples illustrate this concept—one related to electrical noise, one to mechanical noise, and one showing the effects of moving up the sound-quality curve.

Example 1: Electrical Noise — The Ethernet Cable

I was comparing streaming (Ethernet) cables in a system. I know—many people are still on the fence, or firmly on the other side, regarding streaming cables and the entire network's effect on sound quality. However, in my experience, the network side matters significantly. It is extremely noisy, and as soon as we plug an Ethernet cable into our system, that noise finds a pathway in.

In this case, the system had a router positioned next to the audio stand, feeding a streamer via a basic, unshielded Ethernet cable, which in turn connected to the DAC with another inexpensive, unshielded cable. I substituted a well-made, shielded Ethernet cable (with the shield attached only at the source end). I had heard this cable's positive impact in other systems, but in this setup, the improvement was minimal—hardly worth the extra cost. I soon realized the router's noise was overwhelming the cable's effect.

What is the intended benefit of the better Ethernet cable? Its superior shielding protects the signal from external interference. Granted, that is insignificant compared to the internal noise flowing through the cable from the router. Reverting to the basic Ethernet cable, I inserted a high-quality, low-noise switch with a separate linear power supply between the router and streamer. The improvement was striking: a lower noise floor, clearer note decay, more silence between notes, and a stronger sense of rhythm. Once the electrical noise floor of the system was lowered, I reintroduced the better shielded cable—and this time, the result was striking. The sound became even quieter and more dynamic, revealing the cable's true performance.

Example 2: Mechanical Noise — The Power Cable

In a similar experience, I compared power cables in a system. What should a power cord do? At a minimum, it must accomplish two things:

1. Handle the dynamic current demands of the component without restricting power delivery.
2. Address resonant energy entering the component through the electrical system.

We'll focus on the second function, as it represents a significant pathway for mechanical or vibrational noise to enter a system. The setup included a box with an isolation transformer, into which all components were plugged. The transformer's purpose is to lower electrical noise by filtering noise from the incoming AC. I believe it fulfills this promise. Notably, one common trait of all transformers is that they *vibrate.*

It is well known that one of the most effective places to install a power cord is between the wall and the distribution box. Interestingly, in this system, that location produced a smaller improvement than installing a better power cord on the preamp or DAC. Why? Because the power cable in question excelled at managing mechanical energy. The mechanical noise generated by the isolation transformer was more significant than the noise coming from the wall. Thus, installing upgraded power cords *after* the isolation transformer had a much greater effect than installing one before it.

Side note: I encourage isolation transformer manufacturers to devote more attention to the mechanical aspects of their designs. How can transformer vibrations be minimized? How can vibrational energy be absorbed and dissipated effectively? Without careful design, it's easy to achieve zero net gain—or even make the system sound worse. Gains in reducing electrical noise can easily be offset by increases in mechanical noise.

Example 3: Moving Down the Noise Curve — The Clock Cable

The final example concerns the experience of moving down the noise curve and serves as a cautionary tale about drawing premature conclusions. I was evaluating a BNC clock cable in a relatively noisy system. The new candidate cable cost four times more than the existing one. After installation, the music gained a little more body, but most listeners would likely conclude that the modest improvement did not justify the price.

We then installed several noise-reduction devices—on the clock, DAC, network, and AC line. With each addition, the sound improved as we moved down the noise curve. Once all devices were in place, we had reached a level of performance where dynamics were fully realized. Re-evaluating the clock cable at this stage, we found the sound quality dropped noticeably when reverting to the original cable. Reinserting the upgraded cable restored a large, clearly audible improvement—one that most listeners would now consider well worth the cost.

Key Takeaway

The point of these examples is that the same product can seem inconsequential in one system yet transformational in another, depending on where that system sits on the noise curve. Re-evaluate products after making other improvements; you may find their value changes. If something appears well-designed or claims to reduce overall noise, don't lose hope too quickly—its benefits may simply be waiting for a quieter system to be heard.

NOTE: *The audiophile world is filled with widgets that promise to reduce noise of one type or another. Interestingly, one person can claim that a widget works exceptionally well, while another can report that it did not have any perceptible effect. How can both people be right? I tend to believe both people are right. Many widgets work to reduce either electrical or mechanical noise; however, if the system has too much acoustic noise, then the effect of the widget can be overwhelmed. This would result in someone reporting no change in sound.*

Digital Music Reproduction and the Elephant in the Room

Many regard analog music reproduction as the gold standard. Why is this? On paper, digital reproduction is the superior format. Sony famously proclaimed, 'Perfect sound forever.' Although, when the CD was first released in the early-1980s, most audiophiles cried foul—digital sounded dreadful. Problems such as aliasing and jitter were poorly understood at the time. Over the following decades, engineers identified these issues and developed solutions to mitigate or eliminate them.

Still, the sound quality of digital music reproduction continues to lag behind analog. Only within the past five years have digital products emerged that can truly compete with state-of-the-art analog playback. What, then, in the digital domain is wreaking such havoc on the music? I suggest that *noise* is the primary factor preventing digital reproduction from surpassing analog.

What is Included in the Digital Domain?

Consider the various components that make up analog playback and draw a parallel to their digital counterparts. In an analog system, we have a turntable, cartridge, tonearm, phono cable, phono preamp, and, of course, the record itself. In the digital world, the record would be represented by either a CD or a digital file. The turntable corresponds to the CD transport or the server/streamer. The tonearm and cartridge, which pick up and convert the microscopic grooves into a voltage signal, are represented by the DAC in the digital domain. The phono cable's role is mirrored by the digital cable.

But what about the phono preamp? Many consider the phono preamp the heart of an analog system—choose poorly, and the entire experience collapses. Is there an equivalent in the digital world?

I suggest that it isn't a single component; rather, several elements working together to fulfill that role. Each of these components contributes to noise reduction. Examples include low-noise Ethernet switches, linear power supplies, vibration-draining feet, damping plates, grounding devices, clean electrical power, and high-quality, low-noise cables. Each addresses a different noise source, and together they eliminate noise from the digital playback chain. This collective effort can make or break the digital music experience.

Many dismiss these as mere "tweaks," but I consider them essential. This comprehensive approach to noise control is precisely what analog enthusiasts have practiced for decades. Noise must be eliminated. Yet in the digital realm, this fundamental principle is often overlooked or dismissed. Noise is the elephant in the room that few acknowledge—and even fewer truly understand.

Summary: Removing What Is Not Music

The path forward becomes clear when we return to the fundamental nature of music itself. We began this chapter with a quote from W.A. Mozart, who said, "Music is not in the notes, but in the silence between them." This reminds us that if we want music, we must eliminate the noise that fills those spaces. It brings to mind another artist's wisdom—Michelangelo's response when asked how he carved the statue of David: "It's simple. I just removed everything that was not David."

The same principle applies to our pursuit of high fidelity. How do we reveal music through our stereo system? The answer is simple: remove everything that is not music.

In this chapter, we explored the forms noise takes (acoustic, electrical, and mechanical), how it hides inside the ADSR envelope, and how it disrupts timing, tone, and spatial organization long before it becomes audible as hum, hiss, or distortion. The grid model helped visualize this, revealing why small movements in speaker placement or small changes in system noise dramatically alter clarity, coherence, and dynamic expression.

With the concepts of timing, noise, and grid interaction now clearly in view, we can turn to the next question: how should we think about recorded music itself?

Before beginning the hands-on setup process, we must first understand the difference between live and recorded sound, what "accurate reproduction" actually means, and how reference tracks can be used as diagnostic tools for tone, timing, and spatial coherence.

These ideas complete the bridge between the scientific foundations of Section I and the practical precision work that begins in Section II. Chapter 4 takes the final step across that bridge by establishing the listening expectations that will guide the setup process that follows.

RECORD

CHAPTER 4

EXPECTATIONS OF RECORDED MUSIC & REFERENCE TRACK PHILOSOPHY

Before exploring our reference songs, we should discuss recorded music and the broader goals of high-fidelity audio system. Earlier, we established that our aim is to recreate the illusion of live music within our listening space. But what exactly are audiophiles trying to achieve?

The most common answer is *faithful reproduction of the source*—but what that means is widely debated. Two terms often emerge in these discussions: **accuracy** and **musicality**. These ideas are frequently portrayed as belonging to opposing schools of thought—let's call them the *objective* and *subjective* camps. In reality, both must be balanced to achieve great playback.

Understanding this balance between accuracy and musicality gives us the framework we need to evaluate what we hear—and that framework must shape our expectations before we ever move a speaker.

And before we move any speakers, it's important to set the right expectations. Every decision you make during setup—whether you're evaluating timing, tone, or imaging—depends on what you believe recorded music *should* sound like.

This chapter helps establish that baseline.

We'll look at how real, live sound differs from the engineered world of recordings. Once you understand how recordings are created, you can aim for reproduction that feels honest and engaging—not perfect, but convincing.

Then we'll connect accuracy and musicality and show how they work together rather than compete. With expectations sharpened and listening tools in hand, you'll be ready for the hands-on work ahead.

Accuracy vs. Musicality

Defining accuracy in recorded music is difficult. When we attend a live performance—whether in a rock arena, a jazz club, or a symphony hall—we experience musicians performing together in a shared acoustic space. Most recorded music, however, isn't captured this way. It's typically a collection of individually recorded tracks—often performed at different times and locations—meticulously assembled by skilled engineers into what sounds like a cohesive performance.

This isn't a criticism but a simple reality. Studio recordings are often the most polished representations of an artist's vision, but they are rarely "live" performances. And that's perfectly fine—after all, our favorite listening chair at home is far more comfortable than any concert seat, and our sweet spot is always perfectly centered.

Acoustic genres such as symphonic, opera, or jazz typically undergo less manipulation, making their reproduction more about accurately capturing instruments within a real space rather than constructing a mix. Still, even live recordings are influenced by sound engineers—primarily in real time rather than through post-production. Even orchestral recordings are often edited after the fact, with errors replaced by alternate takes recorded in the same venue.

Recognizing how recordings are assembled helps us understand the gap between real acoustic events and what ultimately reaches our speakers. The truth is that most of what we hear in recorded music is a *crafted illusion*, shaped by artistic and technical choices. This raises an important question: Accurate to what?

Answering this requires us to look beyond measurements alone and consider the artistic and technical choices embedded in every recording. This isn't meant as a criticism of the "accuracy" camp but rather as a practical consideration for speaker setup. Likewise, for those in the "musicality" camp, we must ask: *What defines musicality beyond measurement?*

Balancing Objectivity and Emotion

When it comes to accuracy, test tones and measurement tools can only go so far. Even the most advanced modeling of rooms and speakers can't fully capture the nuances of real music. Musical complexity always presents challenges that test signals cannot.

And even if perfect measurement tools did exist, how would we apply them to real-world setups? The interaction between loudspeakers, their associated electronics, and the listening room is far too variable for any measurement alone to provide a definitive answer. Ultimately, a human being must interpret and apply the data to each specific environment.

Evaluating *musicality*, on the other hand, is more like evaluating art—an inherently subjective experience that cannot be quantified. For that reason, true system optimization lies at the intersection of these two ideals: a system that is technically accurate enough to be faithful to the recording, yet musical enough to stir genuine emotion—whether it brings tears to your eyes or makes your pulse quicken.

Framing Expectations

But even a perfectly optimized system is filtered through our own listening history, so our expectations inevitably shape what we perceive. We must also be honest about our expectations. If you rarely attend live performances, your frame of reference will differ greatly from someone who regularly hears unamplified music—or who performs it themselves. The impact of a well-amplified rock concert or the intimacy of sitting in the eleventh row at a symphony hall provides vastly different perspectives on how music should sound and feel compared to even the best home systems.

It's also important to accept that a stereo system can never fully *replicate* the live experience for several reasons. First, most recordings struggle to capture the spatial characteristics of a live event—the "air" and space around the musicians that are essential to a realistic presentation.

Second, live dynamics are fundamentally different. A single flute, for example, produces a coherent sound as one instrument with no internal timing discrepancies. A loudspeaker, by contrast, divides that same instrument's energy among multiple drivers, each responsible for a different frequency range. This introduces inevitable timing and phase challenges. Multiply that by two speakers and an entire ensemble of instruments, and perfect coherence becomes increasingly difficult to achieve.

Despite these obstacles, modern audio systems can come remarkably close. Advances in amplification, power delivery, digital-to-analog conversion, and loudspeaker design have far exceeded most listeners' understanding of how critical proper setup is. The final frontier lies not in better gear but in the skill to align what we already have—to bridge the gap between engineering potential and emotional realism.

Using Live Music as a Reference

Even if orchestral music isn't your preferred genre, attending a live symphony concert is one of the most valuable experiences you can have as a listener. Why? Because it demonstrates dozens of musicians performing together in a single acoustic environment—each occupying a precise spatial location. This serves as a powerful reference for how organized and coherent real sound can be.

It doesn't have to be classical; big band or small jazz ensemble can provide the same insight. What matters is that the performance captures a clearly defined soundstage—a believable sense of space and timing.

Do we know exactly how an orchestra was positioned or how the performance sounded at the moment of recording? Unless we were there, not precisely. But unlike heavily mixed studio productions, live acoustic recordings maintain a natural coherence in both time and space—qualities that provide essential reference points for system tuning.

If your system can convincingly present a live classical or jazz recording as though you were in the hall, the precision required to reach that point will enhance the playback of *all* music, regardless of how it was engineered.

Listening to the Real Thing

Direct exposure to real, unamplified instruments strengthens your internal reference, making it far easier to judge whether a system is reproducing music convincingly. If you've never attended a live orchestral concert, make it a priority. It's one of the best investments you can make in training your ears. Regardless of your musical preferences, a well-recorded live classical performance remains one of the most reliable benchmarks for assessing how effectively your system organizes sound.

When attending, try to get the best seats you can afford—ideally near the center of the Orchestra Seating section. In most American concert halls, rows **G through N** provide a listening perspective similar to how many classical recordings are captured.

Sitting closer to row **G** offers a more immediate and visceral experience, with sound pressure levels often exceeding **110 dB** during climaxes. Moving farther back toward row **N** places more air mass between you and the orchestra, slightly softening the presentation while enhancing the sense of scale.

Keep in mind that microphones used for recording are typically placed much closer to the musicians than the audience. As a result, most recordings convey a "front-of-hall" perspective. When listening, observe the height, width, and depth of the soundstage, and notice how the perception of depth changes depending on your seating position.

Every concert hall has a unique sonic signature, but sitting closer to the performers minimizes the room's influence while still preserving the acoustic environment.

Training Your Ears

Once you begin noticing these characteristics in live settings, you can apply the same awareness to your critical listening at home.

- Pay attention not just to what you *hear* but to how the sound *feels.*
- Ask yourself: Does pinpoint imaging exist in a live setting from your seat?
- How much of the hall's reverberation do you perceive?
- How distinct is the separation between instruments?

These details vary by venue and seating position but analyzing them helps sharpen your listening awareness. Enjoy the performance but try to observe with an analytical ear—as if conducting an experiment.

It can be difficult to balance appreciation and analysis in the moment, but you may find that this deepened awareness enhances both. If you can, attend multiple concerts, varying your seats each time. Every experience will strengthen your reference for how real instruments sound and interact within real spaces—knowledge that will serve you immeasurably in optimizing your own audio system.

Of course, we cannot attend a live concert every time we make a speaker adjustment, so we rely on carefully chosen reference tracks that reveal the same spatial, tonal, and timing cues in a controlled, repeatable way.

Reference Songs

Reference songs are like tools in a toolbox. Each one serves a distinct purpose, offering insight into your system's performance and helping diagnose specific setup issues. Some reveal tonal imbalances; others expose timing or coherence problems. Each also indicates which speaker adjustment—*front-to-back*, *toe-in*, or *rake angle*—is needed to resolve the issue.

More complex tracks can reveal multiple problems simultaneously. When starting out, you may need to listen to a section of music several times to isolate and understand what's happening. As your listening skills mature, the "wrong" elements will become immediately apparent.

Why Use Reference Songs?

Many audiophiles challenge the use of reference songs with a fair question: *"How do you know what this song really sounds like if you weren't there when it was recorded?"*

It's a valid point. Often, you won't know exactly what's correct until you hear it on a properly set-up system. Both *my* perspective and *Stirling's* offer useful answers to this question.

From my perspective, I can discern a great deal through **headphones**. Because headphones eliminate the influence of the room, part of the setup variable is removed. The drivers are also extremely close to the ears, allowing me to hear the left-to-right timing with high precision.

Second, when speakers are correctly positioned, their timing alignment approaches perfection—or as close as is physically possible. Listening to a precisely optimized pair allows me to establish a reliable reference: I know the song sounds *at least that good*. With further refinement, better speakers, or a more optimized setup, I may later discover subtle details I had missed before. Together, careful headphone listening and well-calibrated reference points form a baseline of what a track *should* sound like.

For Stirling, the answer is even simpler: *the correct result sounds obvious*. It isn't about personal taste—it's objective. To illustrate, imagine asking a master chef to identify and cook various doneness levels of a rib-eye steak. While there's room for interpretation, their evaluation is based on skill and experience, not preference. The chef can objectively determine what "medium rare" is, regardless of the steak's quality. Similarly, an experienced listener can identify correct tonal and timing balance by experience and training.

Listening Methodology

I encourage you to listen to all the reference songs using **headphones**. Attempting to judge them through a phone, tablet, or computer speaker is futile. You don't need costly equipment—a decent pair of mid-range headphones works perfectly well. For instance, I often use Bluetooth bone-conduction headphones that cost around **$159,** and they reveal much of what's happening within each track.

If you're unfamiliar with the tone of specific instruments—clarinet, flute, oboe, bassoon, horn, timpani, and others—start with *Prokofiev's Peter and the Wolf*. Listen to the first movement several times until you can easily distinguish the clarinet's woody resonance from the oboe's reedy texture or the bassoon's depth. Building this mental "tonal library" will dramatically improve your ability to recognize instrument characteristics during setup.

How to Use This List

Next are the songs used throughout the speaker setup process. They are listed in approximate order of application. Each entry describes its purpose ("tool"), what to listen for, and the information it provides about your system. As you read and listen, keep notes on how each passage sounds in your setup—use them as checklists for progress.

Mulatu Astatke — "Tezeta (Nostalgia)" from *New York–Addis–London: The Story of Ethiopian Jazz*

Tool: Understanding the concept of noise (not used during setup).

This recording was deliberately chosen because it is noisy. Listen carefully and notice how the noise masks the performance. You should clearly hear how background hiss and artifacts distract from the music.

Andreas Vollenweider — The White Wind/The White Boat from the album *The White Wind*

Tool: Assessing a system's noise level (not used during setup)

This recording opens with the sound of wind. The wind should not sound "whistly." It should be smooth and flowing between and around the speakers. The wind should exhibit large shifts in loudness — sometimes quickly and sometimes gradually. The wind should be a three-dimensional object and not sound like it is a screen stretched between the speakers. Extraneous noise in the system will result in wind that sounds thin, whistly, grainy, and flat.

Sergei Prokofiev — *Peter and the Wolf*, narrated by Sting

Tool: Developing familiarity with orchestral instrument tone (not used during setup).

Peter and the Wolf is an excellent introduction to orchestral timbre. The narrator identifies each instrument and its associated character. Listen repeatedly until you can instantly recognize each instrument's sound. This exercise trains your ear to differentiate tone colors—an essential skill for evaluating system balance later on.

Birdy — "All About You" from the album *Fire Within* (0:00–0:45) (Not Inverted)

Tool: Finding proper *toe-in* angle and starting *rake angle* for the left speaker.

This is a classic studio recording. Birdy's voice occupies a solid position in the midrange. During the first 45 seconds, she sings with a relatively steady tone. Her voice should sound smooth, natural, and free of harsh edges. Common errors include:

- A chesty or overly warm vocal quality.
- A nasal or overly bright timbre.
- Excessive "ch, ch, ch" sibilance on certain words.

Any of these indicate misalignment. When the setup is correct, her tone will sound pure, centered, and effortlessly natural.

Pay close attention to the **relationship between the vocal and the guitar**. The guitar has a syncopated, cyclical rhythm. When the left speaker's *toe-in* and *rake* are dialed in correctly, Birdy's vocal phrasing aligns rhythmically with the guitar's cycle.

Nils Frahm — "Chant" from the album *Solo* (0:00–0:35) (Not Inverted)

Tool: Finding the left speaker's *front-to-back* position where bass is deep, linear, and balanced with an open midrange. "Chant" was recorded live in one take with no overdubs. The piano used—the **Klavins M370**—is extraordinary: it weighs 1.8 tons, stands 12 feet tall, and has strings 10 feet long. This massive wall-mounted piano produces immense tonal depth and resonance.

The recording is very close-mic'd, giving it an intimate, "big" presence rather than the diffuse sound of a concert hall.

The track opens with a solid, lower-register piano chord that extends down to **35 Hz**. The transient attack should be immediate and authoritative—not soft or sluggish. A dull or slow rise time indicates an incorrect speaker position.

Figure 3 demonstrates the **ADSR envelope** (Attack, Decay, Sustain, Release) of the opening chord. The Y-axis represents loudness; the X-axis represents time. Note how sharply defined the initial transient strike is—this precision is critical. Frahm is emphasizing the piano's sheer physicality.

Also listen for a **"bounce"** in the sound immediately after the strike. If the transient attack is too slow or the decay uneven, this subtle bounce will disappear beneath the overhang.

Figure 3: First Note (Chord) of Nils Frahm's "Chant"

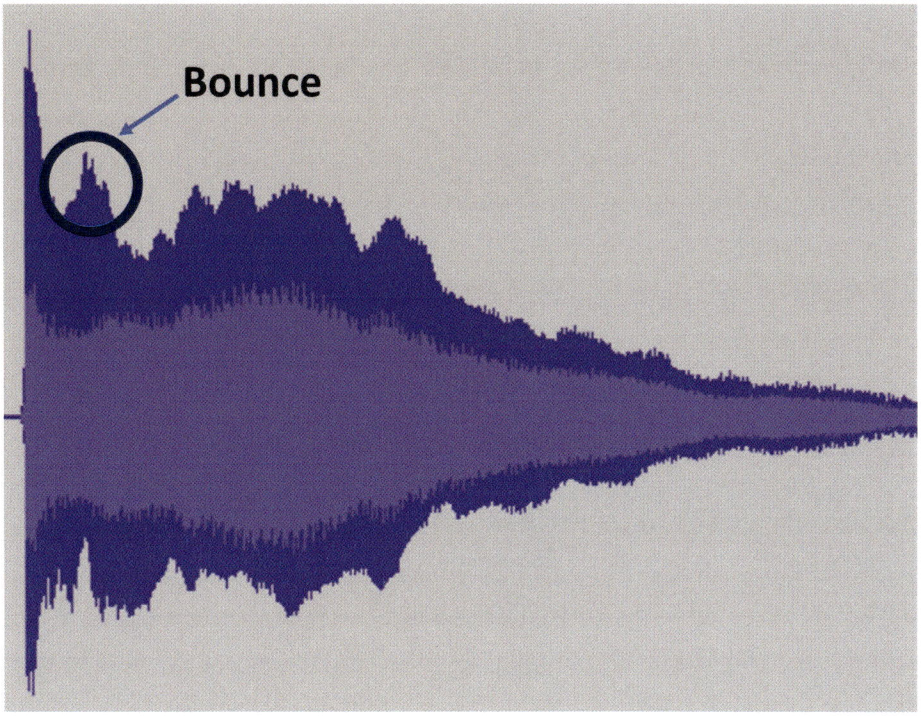

After the initial impulse, the chord should carry a textured "growl." When seated at the listening position, you should feel the bass's energy engage your torso—from waist to neck. You will physically feel the bass. The note's decay should be long and steady, not "breathing" or oscillating unnaturally. When the piano's right hand enters around **0:20**, each keystroke should sound distinct and articulate. You should be able to hear the felt texture of the hammer on the strings with each strike. If it feels as though a blanket covers the upper notes, the midrange clarity is poor. Around **0:28**, listen for a single bell-like high note—it should shimmer cleanly, decay naturally, and reveal how the high frequencies interact with the room.

At approximately **0:32**, Frahm plays a note that decays into silence; the duration and smoothness of this decay indicate how low your system's noise floor truly is. Finally, at **0:35**, a low chord reenters, steady and resonant.

Important: Do not invert this track's polarity. Doing so will invalidate the setup process.

Nils Lofgren — "Bass and Drum Intro" from the album *Live* (0:00–1:22) (Inverted)

Tool: Determining the height of the left speaker to achieve clear and articulate bass.

The microphone feed for this recording is taken directly from the stage speaker—not the bass guitar itself. As a result, the mic captures a real sense of **air pressure** from the vibrating speaker cone. This track produces percussive, visceral air movement, yielding an exceptionally immediate transient. It's an ideal test of your system's ability to reproduce bass **clarity** and **articulation**.

This is a live recording. The track opens with the sound of the audience cheering. The crowd should be heard as a collection of individual voices layered in space—not as an indistinct mass of noise. The audience's energy should sound lively but not piercing. If the applause or screaming sounds excessively bright or edgy—if you instinctively want to turn the volume down or cover your ears—your *toe-in* or *rake angle* is likely incorrect.

Pay special attention to the voice of a man around the **0:04–0:05** mark who shouts, *"Aaaa-ahhh-aaaa-oooh!"* He should sound as though he is sitting well within the audience, not on stage with the musicians. The track's crowd ambience becomes useful later in the setup process, its **primary purpose** is to assess **speaker height**, so focus primarily on the bass lines.

From the start through approximately **0:52**, Nils plays higher-pitched notes. This section is critical for observing **transient attack**. The guitar should never sound slurred—each note must be distinct and well separated. Between **0:12 and 0:24**, listen closely as he runs a quick scale. Each note should start cleanly and decay naturally. From **0:24 to 0:52**, he repeatedly drags his thumb across the bass strings; between **0:44 and 0:52**, you should clearly hear and even *feel* the pluck and release as his finger lets go of each string.

At **0:52**, Nils plays a chord with a subtle *wavering* tone. Listen carefully to how the different frequencies bend. The treble and midrange should not drift independently from the bass. If you hear the higher frequencies bend differently than the lower ones, it indicates **incoherence** in the speaker setup. While this observation alone doesn't always warrant immediate correction, it provides a valuable **read point** for overall coherence.

Around **1:08**, Nils transitions to the lower register and revisits it again at **1:22**. These notes should not sound muddy or thick; they should be well defined, allowing you to hear the texture of the plucked strings. Between **1:22 and 1:26**, he repeatedly plucks the low **E string (41 Hz)**. This passage can easily sound exaggerated or "growly," with the low tone seeming disconnected from the guitar—as if the bass were lying on the floor.

As shown in the illustration, this symptom indicates that the **speaker height is too low**. If you experience this, adjust the speaker by turning **all four spikes** upward **1/8 of a turn** to raise it slightly. Listen again and repeat as needed until the bass note rises in space and locks seamlessly into the body of the guitar.

10,000 Maniacs — "Like the Weather" from *In My Tribe* (0:00–0:19, Inverted)

Tool: Finding the **side-to-side** position of the left speaker with correct **treble**, **midrange**, and **bass** timing.

Only the first ~18 seconds are needed—the repeated introduction before the vocals enter. This passage contains guitars, bass, drum kit, and keyboard layered in a syncopated rhythm that is extraordinarily sensitive to lateral speaker position.

The snare drum provides the timing anchor. Each hit should sound snappy and percussive—a clean crack, not a dull thud or a smeared splash. The bass guitar locks to the snare with a syncopated rhythm that gives the song its groove. When lateral alignment is off, you'll hear the instruments, but they won't be coherent—the snare will seem to arrive at a different time than the bass, and the overall effect is rhythmic confusion.

When the speaker is significantly misaligned laterally, this track sounds almost unrecognizable. The rhythm can feel dull and slow and lethargic or rushed and a bit bright. You may find yourself unable to tap your foot because there's no consistent rhythmic pulse to follow. The instruments occupy the same space but seem disconnected from one another, as though the band members were playing in separate rooms.

As you move the speaker laterally (typically outward), the syncopation begins to emerge. The snare and bass start to find each other. You'll notice the rhythm becoming intelligible—you can start to predict when the next hit will land. The keyboard, which may have been completely buried or phased, becomes plainly audible, occupying its own space.

When the lateral position locks in, the transformation is unmistakable. The groove becomes irresistible—you'll find yourself falling into the rhythm without effort, head nodding or foot tapping involuntarily. The snare has snap and presence. The bass has texture and definition, each note distinct rather than a muddy wash. The keyboard becomes very present and important to the musical line, and you can't believe you couldn't even hear it before. Most importantly, all elements feel unified in time—a single band playing a single song, rather than isolated instruments superimposed.

If you continue moving the speaker laterally past this correct position, the timing will degrade again. The rhythm will become increasingly confused until it sounds nearly as bad as when you started. Keep moving, and you'll find another position where the timing locks in—but this may be a "false zone" where the timing is correct, but the polarity is inverted, producing a strange phase shift. The treble may seem to disappear, or the overall sound may lack body. If you find what seems like correct timing but something sounds hollow or wrong, you've likely landed in an adjacent grid. Move the speaker back toward your original correct position.

The distance between correct and incorrect lateral positions can be remarkably small—sometimes as little as ¼ inch. This is why the track is so useful: its syncopation reveals misalignment that other material might obscure. Once you've found the correct lateral position with this track, mark it. You'll reference this position throughout the rest of the setup.

Megadeth — "Dawn Patrol" from *Rust in Peace* (Non-Inverted)

Tool: Bass and treble **timing** and **texture**.

This track isn't required for the core setup but is extremely helpful if *"Like the Weather"* or *"Bass and Drum Intro"* feel too fast at first. Here, the **bass guitar** and **snare** should land with the correct timing, which requires bass articulation to be audible. Using this in place of *"Like the Weather"* often moves the speaker closer to the correct position—however, the **fragile syncopation** in *"Like the Weather"* remains the more sensitive test of fine timing alignment.

For the first ~**0:30**, the pattern comprises **bass guitar, kick drum,** and **snare**. The song is played slowly—ideal for listeners still developing sensitivity to timing errors. It's in **4/4 time**: the kick typically lands on **beat 1**, the snare on **beat 3**, and the bass guitar plays **each beat**. Ask: *Are David Ellefson and Nick Menza hitting exactly together?* If not, adjustments are needed.

This track also reveals **bass texture**. You should clearly distinguish **kick drum** from **bass guitar** on beat 1, and the bass line should be articulate—without **overhang**. The main notes include **E (41 Hz)**, **D (73 Hz)**, **E (82 Hz)**, and **F (87 Hz)**—a region where many systems smear. As the track continues, Dave Mustaine begins to chant; his low voice can get obscured by the bass if timing and articulation are off. You should **clearly** hear what he's saying about climate change.

Kat Edmonson — "Lucky" from *Way Down Low* (0:00–1:30, Not Inverted)

Tool: Assess **mechanical noise floor**, establish **midrange-to-treble timing**, and refine **toe-in** on both speakers.

"Lucky" opens with **"hoooo'ing"** vocals that should sound **controlled, precise**, and **un-grainy**. This is a superb track for evaluating **resonance control** devices (e.g., footers, damping plates, shelves, racks). A powerful test: isolate components on **rubber feet** (pure isolation), then place them on **energy-draining** supports (e.g., **wood blocks**). Compare Kat's vocal texture—differences can be striking.

Following the intro, note the **vibraphone** strikes. The percussionist uses **double mallets**, so each strike is a **double hit**. The pattern alternates between **8 higher-pitched** strikes and **8 lower-pitched** strikes (one mallet drops in pitch). Each double strike should be **almost perfectly time-aligned**—you should **not** hear a **"ba-da"** effect. Instead, each sound like a single unified **pulse**.

As Kat sings, her voice should be **even** and **smooth**. This recording contains **some inherent sibilance**—you cannot remove it, but correct setup can **clarify** it, so it isn't **sharp** or **fatiguing**. (See the appendix for a detailed explanation of sibilance) Her voice is unusually characteristic; like Birdy, she makes an excellent reference to **nail the correct toe-in angle**.

This track is excellent for exposing issues of **resonance**, **tone**, and **timing**. We're lucky to have *"Lucky."*

The Shins — "Sleeping Lessons" from *Wincing the Night Away* (0:00 — 1:25, Inverted)

Tool: Clarify the left speaker's **rake angle** and **azimuth**.

This studio recording opens with the sound of **someone running upstairs**. On a well-set system, it should **look** (and feel) as if a person is running **up a staircase behind the speakers**. On very high-resolution setups, you may even hear **keys** jingling as they run (suggesting they're about to unlock a door). You should distinguish the **impact transient** of the footfall from the **wooden stair tread thud**.

After the runner reaches the top, the **"bubbles"** begin. On a poorly aligned system, the bubbles blur together and sound **phasey**, disorganized, and even **ceiling-filling** to the point that you'll want to turn the volume down. None of that should happen.

Working with the **left** speaker alone, the bubbles should sound **distinct** and **move in a straight path** up and down the frequency range, from **low → high → low**. When the **bass** enters, it's like adding a **larger bubble** at the bottom of the chain. The more **time-aligned** and **coherent** the speaker becomes, the more **precise** these bubbles sound.

Treat the bubble timing as **quarter notes**. On very high-resolution systems, you can hear **four high-frequency beats** embedded in each bubble—like **sixteenth notes**. Use this to **align the tweeter-to-midrange timing:** the bubble's pulse should land with the **first high-frequency beat**.

This track is also excellent for testing **system tweaks** (e.g., footers, shelves, damping plates). When a tweak **works**, the **resonant bounce** is reduced, and each bubble becomes **clearer**.

Bozzio Levin Stevens — "Duende" from *Black Light Syndrome* (0:00–0:10)

Tool: Fine adjustment of the left speaker's **azimuth**.

The first ten seconds feature a **solo acoustic guitar** played with **high articulation**. Each note should be clearly **separated**, with no **slur** between notes. If notes seem to **tilt** or smear **laterally** (or if the upper harmonics separate from the fundamental), refine **azimuth** in **very small** increments.

Yello — "Till Tomorrow" from *Touch* (0:00–0:45, Inverted)

Tool: Assess and adjust the right speaker's **rake angle**; also reveals **toe-in** and **lateral spacing.**

This track is mostly **computer-generated**, but the **trumpet** is performed by **Till Brönner**. It begins with a centered **electronic bass pulse** you should **feel**—the pressure pulse should hit **dead center**, evenly across the torso. A series of **clicks** and **pops** appears between the speakers: these should be **symmetrical** in both **space** and **time**. The background behaves like an **echo chamber** where these sounds bounce around—similar to the **hall sound** in orchestral recordings.

At **0:23**, **trumpet blasts** begin at the **right**, **grow** in loudness as they move toward the center, then **fade** as they exit on the **left**—all at the **same pitch**. From a stereo-imaging perspective, the trumpets should be **flat**, **equally in focus**, and **evenly spaced** as they traverse the stage. Deviations typically point to **rake** or **toe-in** asymmetries.

CeeLo Green — "Bass Head Jazz" from *CeeLo Green and His Perfect Imperfections* (0:43–1:15, Not Inverted)

Tool: Adjust the right speaker's **rake angle**.

This track uses a **QSound-like** effect—**phasey** and intentionally immersive. Listen from **0:38–0:44**: the **trumpet**, appearing separately on the left and right halves of the stage, helps assess **soundstage level**. This is easier than *"Till Tomorrow"* because the stage extends **far left to far right**.

From **0:46–0:52**, the audible **"breathing"** comes from both left and right speakers, symmetrically about the **centerline**. From **1:03–1:09**, a QSound-style image **circles around your head**. When that circle is **flat**, **even**, and **unbroken**, your **rake angle** is very close to correct.

Ghost Rider — "Make Us Stronger" from *Part of the Dream* (0:00–1:15, Not Inverted)

Tool: Precise **rake, azimuth**, and **front-to-back** adjustment of the right speaker.

This is a very interesting track, as it appears to be totally synthetic. The progressive psytrance Israeli artist Vlad Krivoshein uses wide vistas, and dynamic, pulsing rhythms to create an impressive sonic spectacular. Aside from fairly brief spoken vocals (which could be a real voice or not), the rest of the music never actually "occurred" in space. No microphones were used, and no air was moved to capture these sounds. The stereo panorama, the spatial placement, the sub-bass architecture—all of it was invented on a computer.

So, in some sense, the recording is "perfect." Krivoshein has total control of time. There is no human inflection, hesitation, mistake, or tonal aberration, other than what Krivoshein programmed. To be sure, this is full artistry at play here, and to give him any less credit than any other musician would be wrong. But from the point of view of a human using this cut as a setup tool, there is no naturally occurring reality for the pattern-matching or recognition part of our brain to access.

When using most of the other listed music "tools," we use that music to reconstruct an event. We are trying to compare the fidelity of what we hear to something—a voice in a room, a bow across a string, a physical event that our lived experience tells us should sound a particular way. But with this tool, we are not reconstructing. We are evaluating transmission—whether the system faithfully conveys the producer's spatial and dynamic intentions, even though those intentions were born in software rather than in air. It's the only type of recording that actually has a black background. And what we are evaluating is concrete, even without an acoustic reference to compare against.

A track built around hard "gated" silences and explosive rhythms is a tremendous test of a system's ability to start and stop. A poorly set up system smears those transients and transitions—the silence isn't black, the attack isn't instantaneous. What you are hearing is driver misalignment, room interaction, mechanical instability, and electronic noise.

Krivoshein's wide panning and psychoacoustic spatial effects are a torture test for phase errors between channels. If toe-in or rake is off by even small amounts, those fabricated spatial cues collapse or wander. The fact that the source is artificial doesn't make the phase relationship between two loudspeakers artificial.

Synthesized sub-bass is more demanding than most acoustic bass instruments because it can be spectrally purer and extend lower. It exposes room modes, port resonance, and boundary interactions with nothing to hide behind. There is no harmonic complexity to mask problems. And the structural language of this music — build, silence, explosion — is a stress test of the system's dynamic window. You are hearing whether the quiet passages are truly quiet, and whether the loud passages remain clean and uncompressed.

Presenting an invented spatial reality with conviction — and doing so in a way that is immersive and coherent — demands that every parameter of the system be correct, because there is no familiar reference to cover errors. The system must earn your belief from scratch.

Electronically produced music typically operates on a precise, machine-like rhythm. There is no drift, no human variance in timing. This makes it an exceptionally useful tool for identifying micro-timing errors in your speaker setup—those small misalignments that the more complex textures of acoustic recordings can obscure. This track is not used for the foundational positioning work that the other tools in this process provide. This is a refinement tool. When you are stuck and cannot determine which direction a speaker needs to move, this track can break the tie. Remember, anything you do will change what you just observed, so the key is to be deliberate about what you want to hear.

The First 21 Seconds

What can we learn from 21 seconds of electronic music? Let us examine that. From 0:00 to 0:21, dead center between the speakers, you will hear a repeating rhythmic cell: two quick pulses followed by two spaced pulses—**duh-duh, duh, duh**. The two quick pulses occupy the same duration as each of the single pulses that follow. What matters here is the silence between them.

Against the black background of this recording, each pulse should appear and disappear cleanly, with distinct edges. The gaps between the components of the pattern should be equally black. If the pulses smear or slide into each other, or if the silence between them is not uniform, the system is telling you something about timing coherence at your listening position. Use those gaps. They are the diagnostic signal.

If the space between the notes is at all vague, clarify it with a very minor sideways bump.

If one of the notes gets pulled off center, it means a minor toe-in adjustment should be investigated. I often defocus my listening to hear the two quick pulses and the two spaced pulses as three longer pulses. Taking the focus off the transient and clarity components lets me regroup them as three evenly spaced sounds, as in dah-dah-dah.

Although the first two quick pulses suggest something more frenetic, taken as a group of three the tempo is squarely in the 150 bpm Allegro range. Not exactly slow, but 150 bpm is still well within the brain's ability to distinguish individual notes within a measure. Getting the micro-timing and clarity right at this tempo, where each note is precisely placed and clearly distinguished, pays dividends when you move to more complicated or nuanced music. What you correct here will carry forward to all music.

When the three dah-dah-dah notes feel unevenly spaced, the tempo will either feel sped up or lazy. This is often caused by a transient from one speaker arriving before or after the other speaker. The smallest of bumps forward or backward will bring the timing of the cluster into line.

Quick Listening Checklist (0:00–0:21)

Listen for the four-note pattern: duh-duh, duh, duh

Focus on the **silence between the pulses**.

What to Listen for

- Each pulse should appear and disappear cleanly.
- The silence between pulses should be equally black.
- The two quick pulses should feel tightly grouped.
- The pattern should regroup naturally into three evenly spaced sounds: dah–dah–dah.

What Problems Sound Like

- Pulses smear together → timing blur at the listening position
- Uneven silence between notes → lateral misalignment
- One note pulls off center → toe-in issue
- Three-note regrouping feels rushed or lazy → front/back timing mismatch

Possible Adjustments

- Slight sideways bump
- Minor toe-in adjustment
- Very small forward or backward movement

Denver Jazz Quintet with Ginger Baker — "Cyril Davies" from *Coward of the County* (Inverted)

Tool: Evaluate **left/right perspective** and **timing**.

Here, **trumpet** is on the **left** and **tenor saxophone** on the **right**. The **drum kits** are set back in the stage, with the **hi-hat** emerging from the **right** channel. When setting the **left** speaker, you'll know you're close when you can clearly hear the **sax** standing **next to the right speaker**, even with the **right** speaker turned away. When moving to the **right** speaker, the **sax** should remain clearly **right**, slightly **behind** the plane of the trumpet.

Tweeter timing can be evaluated by comparing the **hi-hat** with the **kick drum**—the cymbal should **land on the beat** in a slightly **syncopated** relationship.

Yim Hok-Man — "Poem of Chinese Drum" from *Master of Chinese Percussion* (Not Inverted)

Tool: Evaluate **deep bass centering** and **tone**, **transient attack**, and **timing**.

Recorded in **one take** with the performers playing together, this track is rich in **timing** information. The **rim-strikes** should sound like **all** drummers are playing precisely **together**. If setup is off, their strikes sound **disorganized** or **random**.

The **big low-frequency drum** is an excellent bass test: it should be **centered** and sound **round**, with a deep **boom** (~**44 Hz**). Beware of setups that overemphasize **harmonics** relative to the **fundamental**, which will **pitch** the drum too **high**.

Around **0:36**, strong **left/right** simultaneous strikes should sound **together**: **"Boom, boom, boom,"** not **"Ba-boom, ba-boom, ba-boom."** Timing won't be mathematically perfect (they're human), but close alignment is the target.

Leonard Cohen — "Treaty" from *You Want It Darker* (0:00–0:50)

Tool: Evaluate **vocal centering** and **upper-bass coherence**.

Cohen's voice sits in the **upper bass** and should image **dead center**, as most studio vocals do. Many systems struggle here: his voice can become **too chesty** and lose **articulation**, making words harder to understand. Almost any studio vocal by Cohen works; this one is ideal because it combines his trademark **gravelly** tone with **breathiness** (higher frequency). As he is **one person** singing into **one microphone**, all vocal components should emanate from the **same spatial location**. Early in right-speaker setup, you may hear the **low** portion centered but the **higher** portion pulled **left (or right)**—a sign of misalignment to correct.

Susie Suh — "Harmony" from *Susie Suh* (0:00-0:33; Not Inverted)

Tool: Evaluate clarity in the **100–200 Hz** range.

This studio track opens with guitar strumming recorded on the **edge of murkiness**. Any blur your system adds in the **100–200 Hz** region becomes immediately apparent. The first line can be hard to understand if this range is smeared—but she is clearly singing, **"I don't know your name."** Use this track to reveal overhang or lower midrange congestion.

Ruth Moody — "Dancin' in the Dark" from *These Wilder Things* (0:00-0:45; Inverted)

Tool: Evaluate **300 Hz** clarity and **centering**.

At the start, a guitar on the **right** and **left** creates a tight rhythmic **bounce** across the stage when the system is aligned. Ruth's voice is expressive yet **never** shouty. If around **0:27** (on the phrase *"Feelin' the same way"*), her voice turns **shouty**, the setup is off. At **0:24** (*"I come home in the morning"*), the **"M"** consonants should **not** smear together. Slight misalignment makes these words **inarticulate**; a large error makes the phrase **aggressive** and **resonant**.

Cécile McLorin Salvant — "La Mal de Vivre" from *For One to Love* (0:00-0:35; Inverted)

Tool: Evaluate **800 Hz** clarity, **ease**, and **centering**.

This piece has a **melancholy** character. When Cécile sings, it should **never** sound **aggressive**. The test comes on the third phrase, **"Ça vient de loin."** It should sound **effortless**, almost as if she's **whispering** to you. As depicted in Figure 4, if the left speaker is not correctly positioned, this phrase can become **edgy** to the point of sounding like she's **stepping out of the soundstage with a machete**. Use this as a clear indicator of precision in the **approximate 800 Hz** band.

Figure 4: "La Mal de Vivre'" Should never sound this way!

Chris Stapleton — "Death Row" from *From a Room: Volume 1* (0:00-1:00; Not Inverted)

Tool: Evaluate **kick drum articulation** and **centering.**

The introduction's **right-to-left timing** is intentionally imperfect. When the speakers are properly aligned, you'll hear a subtle **bounce** between the right and left guitars—creating a rhythmic, organic feel across the soundstage.

This track is primarily used to assess the **kick drum**. The impact should be **solid**, **tight**, and **centered**—hitting you **dead center** in the chest. The drum's fundamental frequency sits around **35 Hz**, distinct from the frequencies tested in *"Till Tomorrow"* or *"Poem of Chinese Drum."* The goal is a deep, coherent pulse that locks into the mix without overhang or bloom.

Sara Bareilles — "Yellow Brick Road" from *Brave Enough: Live* (0:00-0:23; Not Inverted)

Tool: Fine-tune the **right speaker's rake angle.**

In this **live recording**, focus on the first **23 seconds**, during which Sara Bareilles plays a **single chord twenty times.** The piano sound emanates evenly from both speakers, spanning a narrow frequency band.

Imagine a **high-frequency string** stretched tightly between the two speakers. The "string" should appear **level** and **taut**. The left half should not sound higher or lower than the right, and the tonal density—or perceived **thickness**—should be consistent from one side to the other.

This test helps reveal **soundstage height differences** between the left and right speakers. Even minor rake-angle variations can cause one side of the image to rise or fall perceptibly. When the setup is correct, the piano's image will appear **perfectly horizontal**, evenly balanced between the speakers, and stable in height and tonal weight.

(See the following illustrations for visual reference. The first example represents the ideal condition.)

Level and Consistent Tone

Low on The Right

Tone is Thick on the Right

Juilliard String Quartet — *Debussy: String Quartet in G Minor*, Living Stereo (0:00 — 0:04, Not Inverted)

Tool: Fine **right-to-left speaker alignment**

This is **Debussy's only string quartet**, a masterwork of texture and color. For setup purposes, we will focus only on the **first four notes** of the **second movement**—though I encourage you to listen to the entire piece once your system is optimized.

This movement begins with a **pizzicato** passage—each note plucked rather than bowed. The tempo is moderate, allowing the **attack**, **release**, and **space** between the notes to be clearly audible from both speakers. The Living Stereo recording was chosen specifically for the **evenness** and **clarity** of this opening pizzicato; other performances vary too much in timing and tone for calibration purposes.

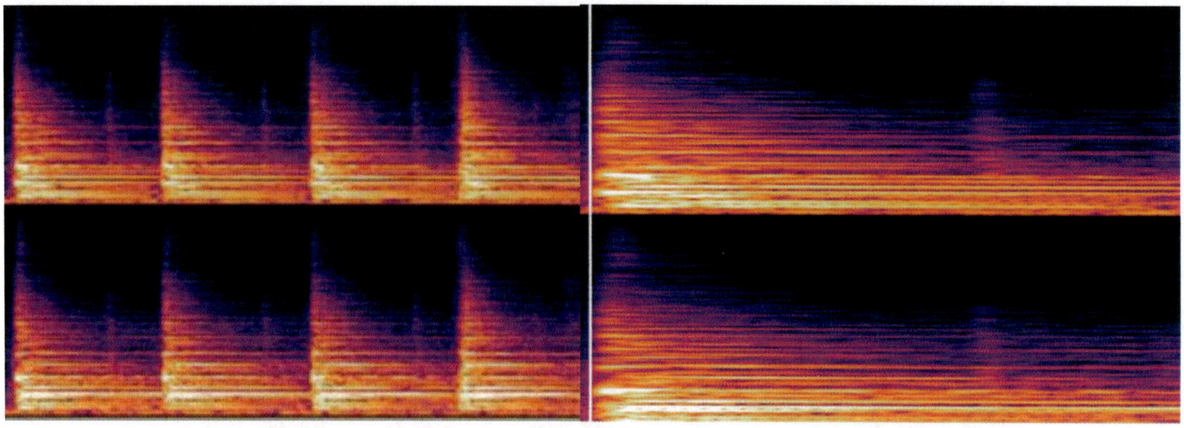

Focus your attention on the **attack** of each note. You should hear the strings' **pluck** occur **simultaneously** from both speakers—cleanly and evenly spaced. When the system is properly aligned, you will hear a steady, consistent pattern:

pluck, pluck, pluck, pluck

If alignment is off, you'll instead hear something like:

pa-pluck, pa-pluck, pa-pluck, pa-pluck

—indicating that one channel is slightly delayed relative to the other.

On the previous page are spectrograms of the first four notes, along with a zoomed-in view of the first note. The **top** graph represents the **left channel**, and the **bottom** represents the **right channel**. Notice how closely the two match—both in timing and amplitude. The zoomed spectrogram on the right shows the first note's **perfect synchronization** between channels.

When your system reaches this level of precision, the quartet's sense of **cohesion**, **placement**, and **attack timing** will feel effortless—an unmistakable sign that your right and left speakers are working in perfect harmony.

London Philharmonic Orchestra — Shostakovich Symphony No. 1, Allegretto, Kurt Masur (SACD Not Inverted, CD Inverted)

Tool: General-purpose track — can be used to adjust **all setup parameters**

This is an **exceptionally well-recorded** piece of music, notable for its **huge dynamic range—Roon reports a dynamic score of 22.** In Roon, "scoring 22" refers to the track's **Dynamic Range (DR) score**, which is a numerical estimate of how much variation there is between the quietest and loudest parts of a recording. For setup purposes, you will need to turn the volume up quite high. The orchestra remains quiet until approximately **4:42**, when it finally plays at **full intensity**.

If you're concerned about your speakers' **dynamic handling capability**, pause or lower the volume before reaching that section.

This recording offers a vivid **palette of orchestral color**. Every instrument should sound **natural** and **believable**:

- The **muted trumpet** should sound like a trumpet played with a mute—clear and focused but softened.
- The **clarinets** should have a distinctly **woody** timbre.

- The **flute** should sound **smooth** and **airy** yet still maintain **body**.
- The **bass drum** should feel **concussive**, with a sense of deep pressure rather than a simple "thud."
- The **violins** should sound like **real strings**—never **etched** or **screechy**.

If anything in the orchestral texture sounds **unnatural**—too bright, too dull, too distant, or too forward—it indicates a setup issue.

Listening for the Sound of the Hall

Many orchestral recordings begin with several seconds of "silence" before the first note, as the engineer presses *Record* and the conductor raises the baton. In reality, this is **not silence**—it is the **sound of the concert hall itself** slowly emerging into audibility.

In *Shostakovich's Symphony No. 1*, you can hear the hall's ambient noise fade in during the first few seconds before the **trumpet enters**.

Figure 5 is a **spectral representation** of the "sound of the hall":

- The **X-axis** represents **time**, and the **Y-axis** represents **frequency**.
- The **top** trace shows the **left channel**, and the **bottom** trace shows the **right**.
- **Brighter colors** indicate higher amplitude (louder sound).

After pressing play, there is roughly **0.4 seconds** of true silence (represented as black in the spectrogram). Then, a gradual **ramp-up** of hall noise occurs over about **0.7 seconds**, followed by roughly **2.8 seconds** of steady ambient sound before the first trumpet note is played.

This short prelude contains a wealth of valuable information. Listening carefully to this section will help you assess:

- The **room's contribution** to perceived space and decay.
- The **balance and symmetry** between the left and right speakers.
- The **speed** at which the hall ambience "rises" (which reflects front-to-back alignment and transient response).

A properly set-up system will present the **hall sound** as a three-dimensional, evenly distributed field—spanning from **floor to ceiling** and **wall to wall**—before the orchestra begins.

Figure 5: Spectrogram of Hall Sound in *Shostakovich's Symphony No. 1*

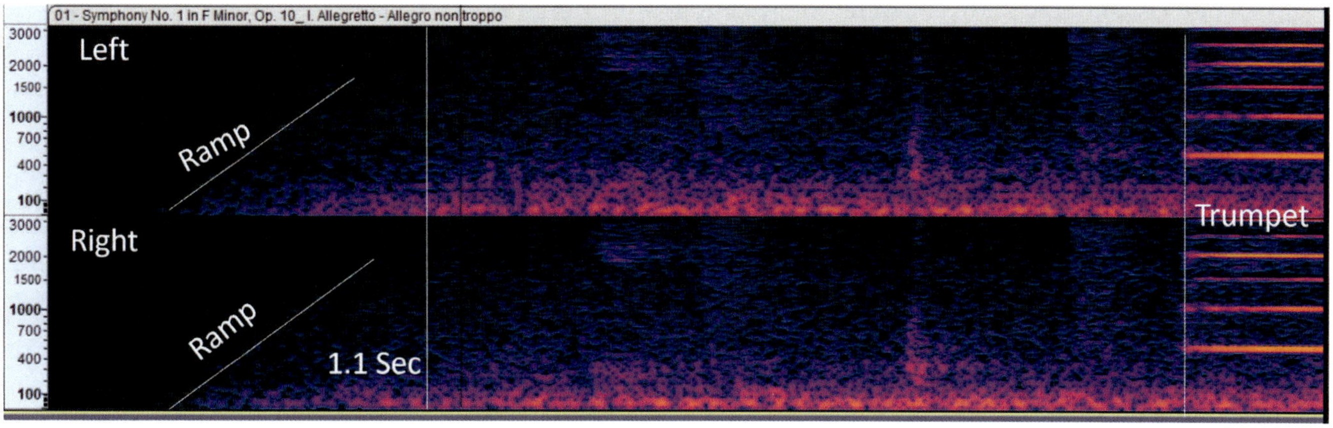

Notice in the **spectrogram** that the sound does not transition instantaneously from complete silence to a fully developed hall ambiance. Instead, there is a **gradual ramp-up** that takes approximately **0.7 seconds** for the hall sound to reach full presence.

This **ramp rate** is a critical observation. When listening, the hall should appear quickly and naturally—*as though the space suddenly comes alive.* If you perceive the hall "coming up" sluggishly or unevenly, something is likely wrong with your speaker alignment. The most probable cause is an incorrect **front-to-back position**.

When this happens, return to *"Chant"* and confirm that the **initial piano strike** sounds **solid and planted,** with a distinct **growl** and **no breathing** or wavering in the low frequencies.

Analyzing the Spectrogram

Also, observe the **thickness of the low-frequency band** in the spectrogram. This visual cue reveals the immense amount of **low-frequency energy** present in nearly all orchestral recordings. At the same time, notice how much **high-frequency information** is distributed throughout the sound field.

Play the **first four seconds** of the recording several times. Turn the volume higher than usual—remember, we're not yet listening for musical content but for spatial information. Don't worry about stopping playback before the trumpet enters; that initial note is relatively quiet.

As you listen, ask yourself the following:

- Do I hear **anything at all** during the ramp-up?
- Is there a **low rumble** or an almost imperceptible **air movement**?
- Are there **tiny clicking or shifting sounds** that suggest the hall coming alive?

- Where do I perceive the **hall sound**—left, right, below, above, or beyond the speakers?
- Does the sound seem to **extend to the ceiling or beyond the side walls**?

The "Sparkly Mist" Analogy

This prelude is one of the most effective tools for evaluating **soundstage structure**. If you imagine the hall's acoustic signature as a kind of **sparkly mist**, it should **fill the room**—extending **wall to wall** and **floor to ceiling.**

When the setup is slightly off, this mist may rise only to **mid-speaker height**, cling to **one side** of the room, or form **holes** in the soundstage where the ambience disappears. Any of these symptoms indicate a setup issue.

When adjusting the **left channel**, for instance, if you perceive **no mist** or ambient extension on the **right side** of the room, that typically points to an error in the **front-to-back** and/or **lateral positioning** of the left speaker.

The goal is to achieve a **smooth, continuous, and enveloping field**—a hall sound that feels *alive, uniform, and immersive*, seamlessly surrounding the listener from all directions.

Figure 6: A Depiction of the Sound of a Hall

At approximately **five seconds**, the trumpet enters and should appear positioned in the far-right rear of the soundstage as seen in Figure 6. (Farther back would be even better.) This placement is a **key indicator** when fine-tuning the left speaker.

If the trumpet does **not** audibly anchor itself in that **right-rear location**, the **left speaker** is **not** yet correctly positioned. Adjust its placement incrementally until the image locks in space.

When this is correct, the experience can feel slightly uncanny—the **right speaker**, though turned **away** from the listening position, will still project the **trumpet's sound** from that distant right-rear corner of the hall. This phenomenon confirms that the acoustic timing and spatial coherence between the two channels have been achieved.

Figure 7: *Shostakovich's Symphony No. 1* — Frequency Spectrum of Trumpet's First Note

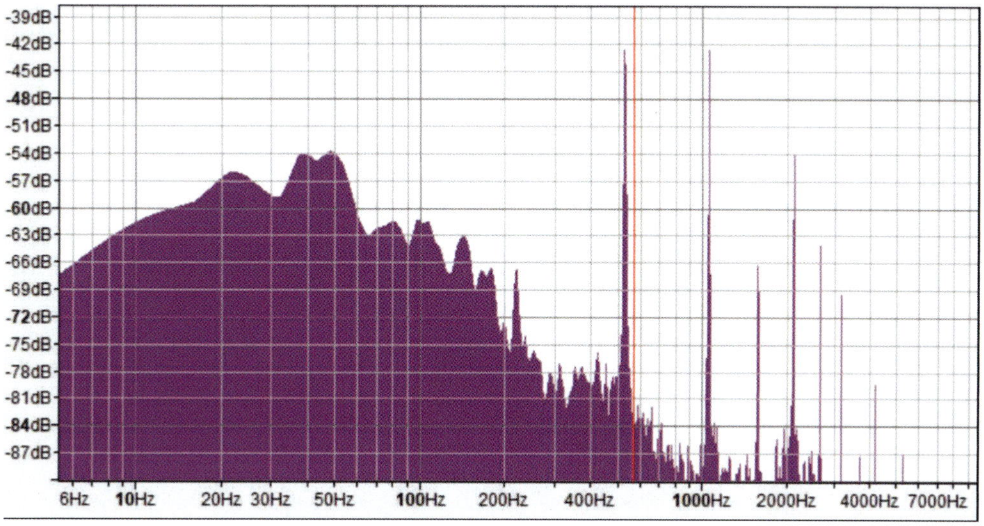

The **graph** in Figure 7 shows the **frequency spectrum** of the initial trumpet tone. There are several key observations to note.

First, notice how much **low-frequency information** is present—even though only the trumpet is playing. This low-end energy contributes **significantly** to our **perception of space.**

Second, observe the **numerous frequency spikes** visible in the spectrum. The note played is **C5**, with a **fundamental frequency** of **523 Hz**. To the right, you can see evenly spaced **harmonic frequency spikes**—the second harmonic at **1,050 Hz**, the third at **1,577 Hz**, the fourth at **2,102 Hz**, and the fifth at **2,627 Hz**.

Importantly, the **fifth harmonic** is only about **12 dB down** relative to the fundamental, indicating that the higher harmonics carry substantial energy. This matters because the harmonic structure is reproduced by different speaker drivers. The **fundamental** and **first few harmonics** likely originate from the **midrange driver**, while the **fourth harmonic** falls within the **tweeter's** operating range.

The **spatial placement** of the trumpet in the soundstage depends on the **timing and alignment relationship** between the tweeter and your ears. Subtle timing errors here can shift or blur the perceived position of the trumpet.

Other Key Moments from Shostakovich and Kurt Masur

From **0:14 to 0:18**, the **oboe**, **clarinet**, and **flute** play rhythmically together. You should be able to clearly distinguish the **individual timbre** of each instrument. The passage should sound **natural** and **balanced**—never bright, edgy, or congested. Avoid a "blobby" high-frequency mass; you should hear distinct instruments occupying defined spaces.

From **0:29 to 0:33**, both sides of the orchestra play **ascending and descending scales**. The instruments should remain **separated yet connected** in the center of the stage. The interplay should sound **delicate and synchronized**. Because of the stereo microphone technique used, there is a **subtle mistiming cue**—you'll hear the **left side "land" slightly before the right** at the bottom of the scale. This is natural and should not be corrected but keep it in mind when judging speaker timing relationships.

Throughout the **introductory section**, various **woodwinds** and **strings** interact. Pay careful attention to the **center image**. If too much **grid overlap** occurs, the sound will feel **congested** or **chaotic**. Conversely, if the grids are pulled too far apart, a **hole** may develop in the middle of the soundstage. The goal is to **"stitch" the two sides together** into a cohesive whole. Correcting this often requires **small lateral bumps**—usually one or two slight adjustments.

At **0:56 to 0:57**, two **pairs of bass drum strikes** occur (each drum hit twice). The second pair is noticeably lower in tone than the first. Both pairs should sound **perfectly centered**, with solid impact and no lateral drift.

At **1:13**, the **piccolo flute** plays a high **G6 (1,568 Hz)**—an exclamation point in the musical line. The note should not sound **edgy** or **shrill**; the flute must retain **body** and remain **integrated** with the ensemble.

From **2:25 to 2:26**, there is a brief **dynamic swell**—a moment of crescendo that should feel smooth, not abrupt nor compressed.

Between **2:33 and 2:41**, the **bass section** is **plucked**, producing distinct notes separated by strong silences. Each pluck should create a clear **pressure pulse** in the room, without **overhang** bleeding into the next note. The silence between notes is just as important as the sound itself.

From **2:42 to 2:52**, the **flute solo** should sound **smooth**, with all notes emanating from a **single, stable point in space**. The flutist tends to meander slightly, and the topmost note may sound **disconnected** if alignment is off. Ensure the flutist stays **"seated"**—the image should not drift toward the violin section.

At **3:36**, two **violinists** play alternating halves of a solo, both belonging to the **first violin section**—the **second chair** takes the first phrase, and the **first chair** takes the second. Their **timbres differ** slightly, so don't attempt to "correct" this with speaker adjustments. Because both violins sit on the **left side** of the orchestra, **both microphones** pick up their sound, introducing a **slight echo-like delay** in the **right channel**.

The violin and its **reflected shadow** should sound **together**, maintaining a natural cohesion. For the most precise fine-tuning—especially late in setup—focus on **matching the decay** of the note at **3:52** between the two channels. Achieving this alignment brings an exceptional sense of realism and balance to the entire soundstage.

***Shostakovich's Symphony No. 1* Violin Solo**

The orchestra concludes during the final **20 seconds** of the first movement, followed by the subtle sounds of the microphones being lifted. In these moments, you can hear the musicians **rustling** on stage, adjusting their chairs, and even **coughing softly**.

More importantly, you can now hear the pure acoustic signature of the hall—its natural reverberation and lingering decay. This ambient information is extremely valuable during the final stages of setup. Use it to verify that the **hall sound** remains **even, balanced, and consistent** across the entire soundstage.

When both speakers are perfectly aligned, the sense of space will remain **stable and uniform** from left to right, with the reverberant field gently enveloping the listener—just as it would in the concert hall itself.

Orpheus Chamber Orchestra — Rossini Overtures, *L'Italiana in Algeri* (0:00-1:50; Inverted)

Tool: Evaluate **bass, midrange**, and **treble timing** for both left and right speakers; assess **noise floor**.

L'Italiana in Algeri—or *The Italian Girl in Algiers*—is full of contrasts: **quiet vs. loud**, **slow vs. fast**, **light vs. heavy**, **mischievous vs. regal**. The music contains a wealth of information for setup refinement.

The **hall ambience** is audible within the first few seconds of the recording, though it is less prominent here than in *Shostakovich's First Symphony*. Extracting the hall sound will require **careful listening** and **fine adjustments**. As the instruments begin, you can hear the subtle reflections from the room—reverberations that seem to **circle around your head** when the system is properly aligned.

The Opening Pizzicato Section (0:03–0:23)

As the music begins, all the strings—**violins**, **violas**, **cellos**, and **basses**—play **pizzicato**, meaning the strings are **plucked**. Each pluck between **0:03 and 0:23** should exhibit a **crisp attack** followed by a **long, natural release**.

The score shows an **eighth note** followed by an **eighth rest**, providing ample space for silence between each pluck. When the setup is correct, this passage feels **playful** and **sneaky**—as though someone were tiptoeing toward their unsuspecting victim.

Between **0:16 and 0:20**, a gentle **dynamic swell** occurs as the plucking becomes slightly louder. Immediately after, the volume drops, creating contrast with the sudden **orchestral crescendo** at **0:24**—a classic Rossini "surprise."

This passage is a powerful test of **dynamic contrast**—how effectively your system conveys the gradations between **piano (p)**, **pianissimo (pp)**, and **pianississimo (ppp)**—while maintaining a sense of **lightness** and **suspense**.

A Note for the Detail-Oriented Listener

At approximately **0:14**, you'll hear a **squeaky chair** on stage—likely a violist shifting position. While not musically important, this small detail is revealing. As your **left speaker** becomes more accurately positioned, the chair's squeak will acquire **shape and body**, transforming from a flat sound into a defined physical presence within the soundstage.

With extremely fine adjustments, you can even **modulate** this squeak—bringing it forward, backward, or losing it entirely within the "grid." This is a wonderful, practical demonstration of the **grid alignment concept** in action.

The Oboe Solo (0:29–0:43)

At **0:29**, the **oboe** enters, taking center stage (although positioned left of center) above the softly plucked strings. The solo should stand **clearly forward** of the ensemble, with the strings continuing their rhythmic pizzicato in the background. The oboe must never sound as though the orchestra is competing for space—it should feel **effortlessly prominent**, yet naturally integrated.

When reproduced correctly, this passage exemplifies **musicality**—expressive, emotional, and dynamically alive. The oboist plays with exquisite **vibrato** and subtle **dynamic shading** that communicate both precision and feeling.

At **0:36**, the oboe sustains a **G4 (392 Hz)**—the lowest tone played during the solo. This tone reveals how much **body** the oboe has in your setup. If the sound feels thin or the image seems "elevated," the **speaker positioning** may be pitching the instrument too high.

At **0:43**, the oboe holds its highest note, **C6 (1,046.5 Hz)**, with radiant tone and controlled vibrato. If the system is misaligned, the **vibrato** will sound artificial or uneven. When the setup is right, the note will sing—fluid, balanced, and emotionally resonant.

Timing Relationships

Just as important as tone is **timing**. Pay attention to the **relationship between the oboe and the plucked bass.** The two must be in **perfect synchronization**—their rhythmic interplay defines the musical tension of this passage.

If the **bass notes arrive early or late** relative to the oboe, the result is **time smearing**—a subtle but damaging effect that drains the performance of its emotional weight. When the alignment is correct, the timing "locks in," and the solo becomes both **musically expressive** and **spatially precise**—a hallmark of a well-optimized system.

Oboe Solo in *L'Italiana in Algeri*

Starting at **1:39**, the **orchestra** plays a quiet series of **three notes**. The delay between each note should feel extended—creating a **dramatic pause** that builds **tension and anticipation**. You should sense the air being drawn out of the room, as if in a slow, deliberate countdown:

Three... two... one...

If your system exhibits **excess noise** or **bass overhang**, this brief moment will lose its tension. The silence between notes should feel heavy and alive with anticipation; noise or smear will flatten the emotional impact entirely.

At **1:24**, the **oboe** introduces a **melodic line**, followed at **1:32** by a **clarinet** echoing the same phrase. The clarinet can sound **"quacky"** or hollow if the setup isn't correct. You should hear the clarinet's tone as smooth and articulate, with its natural **woody resonance** preserved—free from smear or glare.

Think back to the earlier oboe solo and how clear and emotionally expressive it was. The clarinet should sound equally musical. If you find yourself thinking, *"Who's strangling that clarinet?"* the **speaker position** likely needs fine adjustment. A small **azimuth correction** or a minor change in **rake angle** (wrong grid spacing) typically resolves the issue.

Beethoven — Piano Concerto No. 1, Allegro con Brio, Arturo Benedetti Michelangeli (Not Inverted)

Tool: Used for both **left and right speaker** adjustments — **lateral, front-to-back**, and **rake.**

Like *Shostakovich's First Symphony*, pressing play reveals the **hall sound** before the orchestra begins. The hall ambience should rise **quickly** and sound **fresh and alive**—not sluggish or dull.

When the music starts, the **strings** enter together: **violins** on the **left**, **basses** and **cellos** on the **right**. This passage offers valuable **timing information**:

Use the **cellos and basses** to fine-tune the **right speaker**.

Use the **violins and cellos** together to confirm **left-to-right balance**.

Both sides of the orchestra should sound **level**, as all instruments share the same stage. For example, the violins should not appear to be playing from a **balcony** while the cellos remain **on the floor**—that indicates vertical misalignment or uneven rake.

The Piano Enters (3:13–3:52)

The next checkpoint occurs when the **piano** enters, just after the cadence concluding the first orchestral exposition at approximately **3:13**. Each note should sound **distinct**, with clear **space** between them—no slurring or smearing.

From **3:30 to 3:35**, Arturo Benedetti Michelangeli plays a series of **triplets**. These must sound **crisp** and **bell-like**, with a precisely defined transient and pure tonal decay.

At **3:40**, a cascade of **descending arpeggios** follows. When adjusting the **left speaker**, the piano should appear to **extend outward** from that side, moving diagonally toward the **center**—with **high notes** on the **left** and **lower registers** moving toward the **right**.

When working with the **right speaker**, the spatial impression becomes subtler. A properly aligned piano will sit **diagonally across the soundstage**, with **high registers** to the **right** and **low registers** to the **left**, just as a real grand piano would appear from the audience perspective.

The Fast Orchestral Section (6:55–7:30)

From **6:55 to 7:30**, the **orchestra** plays a rapid, rhythmically intricate passage. If timing alignment is off, this section will sound **noisy** or **acoustically smeared**, with instruments overlapping incoherently.

When the system is properly time-aligned, each instrument occupies a **defined space**, and the interplay remains clean and intelligible. To correct timing issues in this section, make **small lateral movements** of the **right speaker** until the clarity locks in.

The Oboe Solo and Piano Scale (7:32–9:07)

At **7:32**, an **oboe** solo enters—projected **forward** in the soundstage. The tone should sound **bold yet natural,** never **aggressive** or **squawky**.

At **9:07**, the **piano** plays a series of **high-to-low scales** three times. Each note should be **distinct**, **evenly spaced**, and **precise**, with no slurring between them. While every note must remain separate, they should still feel **connected**—like a single flowing gesture across the keyboard.

This section is an excellent test of **pitch clarity, attack definition**, and the **spatial geometry** of the piano image across the soundstage.

Schubert — Piano Sonata No. 20, Andantino, Leif Ove Andsnes & Ian Bostridge (0:00–7:50, Not Inverted)

Tool: Final polishing — **refining tone, spatial coherence**, and **emotional engagement**.

Azimuth, Lateral, and Rake Adjustments

Franz Schubert composed his *Piano Sonata No. 20 in A Major, D. 959* while suffering from the long-term effects of syphilis, completing it just two months before his death in 1828 at the age of 31. The *Andantino* movement—slow, introspective, and emotionally charged— is written in F-sharp minor (The relative key of A-major), and it unmistakably **pulls at the heart**.

The structure of the Andantino follows an **A–B–A ternary form**. The **first theme** (opening to **2:52**) conveys a sense of **deep contemplation**—quiet, reflective, and searching. The **second theme** introduces feelings of **panic and anger**, before returning to a **modified first theme** at **5:45**, bringing a sense of **acceptance and closure**.

You can almost imagine Schubert seated at his piano one solitary afternoon, pouring out his anguish and fatigue through the keys—every note carrying the weight of his years of illness.

Why This Piece Matters

You might wonder why this historical and emotional background is relevant to speaker setup. The answer is simple: this is the **final test**—the piece that reveals whether all your efforts have achieved true musical and emotional engagement.

Once your setup refinements are complete, sit back, take a deep breath, and press play at the beginning of Schubert's *Andantino*. Listen **without analysis**. Ask yourself:

- *Does the music pull me in?*
- *Can I feel Schubert's struggle, sorrow, and resilience?*

If the answer is yes, your system is aligned—not just technically, but emotionally.

Why the Piano is so Challenging

The **piano** is one of the most demanding instruments for a stereo system to reproduce convincingly. It is a **full-range** instrument, physically large and harmonically complex, producing both **percussive transients** and **sustained overtones**.

This combination exposes weaknesses in **driver-to-driver** and **speaker-to-speaker coherence** more than almost any other instrument.

This particular recording—performed by **Leif Ove Andsnes** and **Ian Bostridge**—is both **beautifully played** and **superbly recorded**. The slow tempo and crystalline recording quality make it a perfect test of system integration. When properly aligned, this piece can be **emotionally moving and immersive**; when not, it will sound **flat**, **thin**, or **mechanically disjointed**.

Listening and Fine Adjustments

As the sonata unfolds, Schubert moves through **dyads and chords** that gradually ascend the keyboard. Each tone should sound **pure**, free of artificial resonances or coloration.

The subtle variations among the dyads provide a means to detect where **driver overlap** occurs—revealing areas where one driver is contributing too much energy relative to another. These cues guide **fine azimuth and rake adjustments**.

By the time you reach this stage of the process, the **lower notes** should already sound rich and balanced. However, as Andsnes ascends the scale, you may find a point where the **purity of tone begins to falter**. Continue making micro-adjustments to the **speaker's attitude**—its rake and azimuth—to preserve tonal coherence all the way to the upper registers.

I call this process **"(w)ringing out the highs."** It's where the final refinements take place, and the last traces of distortion or imbalance are eliminated.

The Repetitive G6 Note

At approximately **3:42**, listen for a **repetitive G6 note (1,568 Hz)**—the familiar *ding, ding, ding, ding*. This note should sound **controlled**, like a **piano key connected to the entire instrument,** not like an isolated frequency spike.

It must **blend naturally** with the rest of the piano—neither poking sharply forward nor sounding out of proportion. When correct, it will possess an **"ivory" quality**—smooth, tactile, and harmonically integrated.

Correcting this requires **very fine driver alignment**—specifically, **minute azimuth** and **rake** adjustments. Even subtle movements can transform this note from piercing or disconnected into something cohesive and lifelike.

When you reach that point—where every tone feels joined to the instrument and the music itself breathes—you'll know the system is not only aligned but **alive**.

Schubert: ding, ding, ding, ...

From **3:55 to 3:57**, there is a **rapid run up and down the piano**. Every note should sound **distinct and articulate**, with **no smearing or blurring** between tones. If you hear any smear or loss of clarity, make a **slight lateral adjustment** to the **right speaker** until each note becomes sharply defined.

Several **piano crescendos** occur at **4:07**, **4:14**, **4:25**, and **4:32**. These moments should sound **controlled** and **dynamic**, not like a **crash** or uncontrolled surge. If they sound overblown or chaotic, move the **right speaker laterally outward** in very small increments until the crescendos gain balance and composure.

It's often most effective to **begin with this section** of the sonata and work backward to the earlier piano run. Achieving clean, coherent crescendos is a broader adjustment, while refining the fast run demands **more nuanced fine-tuning** once the overall alignment is close to ideal.

The Final Note: An Extraordinary Decay

The **final note** of this piece is a marvel. After the pianist strikes the last key, the tone **decays for nearly ten seconds** before fading into silence. When properly reproduced, this moment feels almost **suspended in time.**

As the note diminishes, you should perceive a sense of **pressure releasing from the room**—a physical relaxation, as if the piano and air have been holding tension that finally releases. The **difference in air pressure** is subtle but perceptible through your body rather than your ears.

Approximately **six seconds** after the sound completely fades, the **recording stops**. The entire sequence resides **deep within the noise floor**, revealing how much pressure and ambient energy linger in the listening space.

This passage provides one of the most extraordinary demonstrations of a **low-noise system**. You cannot experience this effect through headphones—it exists only when a full system and room act as one coherent instrument. When the setup is correct, this final note will remind you that silence itself can feel like music.

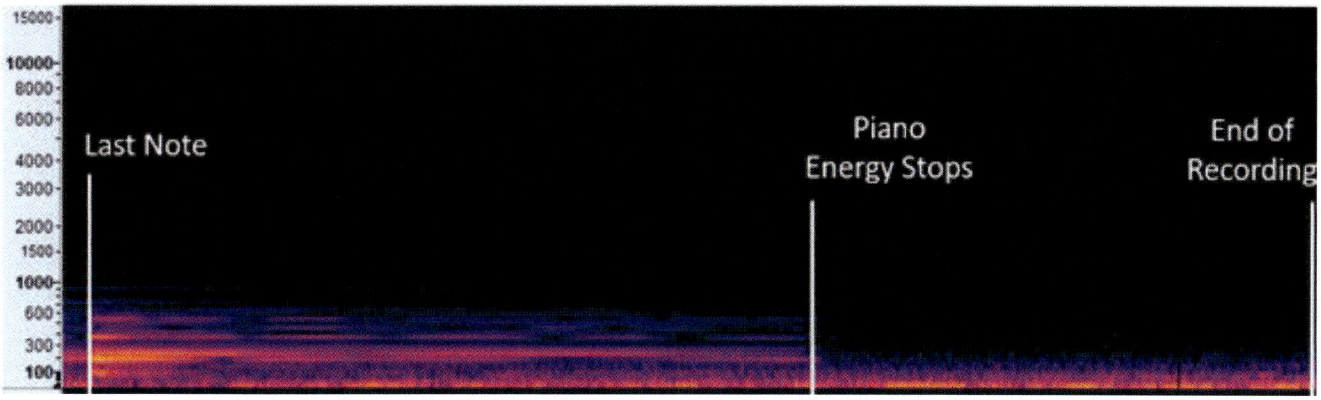

James Horner — "Cosmo… Old Friend" from the *Sneakers* Soundtrack (Non-Inverted)

Tool: Subwoofer or **bass tower integration**

"Cosmo… Old Friend" is recorded at an unusually low level, so you'll need to **raise the volume significantly** for this test—typically about **10 dB higher** than your normal listening level for music such as *Ruth Moody*.

The track features a **slow, steady bass drum beat** that continues throughout. The drum should sound **centered** and positioned **deep in the back of the soundstage**, providing a foundation that anchors the entire recording.

Turn up the volume until the **bass drum** is clearly audible but **avoid pushing the system to distortion or strain**—if your speakers begin to sound stressed, lower the volume immediately to prevent damage.

As the piece unfolds, **woodwind instruments** gradually enter. There are moments where the **downbeat of the bass drum** and the **woodwinds' attack** should land **exactly together** in time. Identifying these precise moments is easiest when listening through **headphones**. Once recognized, use them to fine-tune **subwoofer timing** and **phase alignment**, ensuring both tonal integration and rhythmic cohesion between the low-frequency and mid-range elements.

Dallas Symphony Orchestra — Shostakovich Symphony No. 8, *Allegro non troppo* and *Largo*, Andrew Litton (Inverted)

Tool: Subwoofer or **bass tower integration**

This powerful recording from the **Dallas Symphony Orchestra** is another essential test for subwoofer alignment and integration.

At the end of the **third movement (Allegro non troppo)**, the **timpani** are struck with **immense intensity**, punctuated by **blaring horns** and ending with a sustained **snare drum roll** that transitions directly into the **fourth movement (Largo)**—performed **attacca** (without pause).

At the opening of the **Largo**, you'll hear **three massive "bombs"**—deep, concussive percussion strikes that introduce the movement's desolate landscape. These "bombs" should feel **physical and centered**, resonating through the room without overhang or muddiness.

To approach perfect integration, start by **aligning the slower elements** (the large timpani strikes and sustained notes) before addressing **faster passages**. The goal is for the **"bombs"** to sound **tight, concussive, and spatially coherent**, while the **timpani rhythms** remain **clear and well-defined.**

When properly aligned, the transition from the explosive power of the *Allegro non troppo* to the haunting emptiness of the *Largo* will sound seamless—an emotional and technical benchmark that confirms your **subwoofer integration** is close to perfection.

Necessary Tools for Practical Applications

You now have the listening framework, expectations, and reference tracks that will guide every step of the setup process. The next step is practical: assembling the tools that allow you to convert what you *hear* into precise, controlled movements. These tools make each adjustment—from broad speaker placement to the smallest final refinement—smooth, predictable, and repeatable.

Your Ears

Luckily, these come **preinstalled**—though a little calibration may be required. Chapter 4 helped train them to recognize tone, timing, and spatial accuracy. Now, they'll become your most powerful measurement instrument.

Sliders

Use at least **eight sliders**—four for each speaker—under the spikes to allow smooth movement during setup. The **type of slider** matters: spikes will puncture soft plastic unless a footer is placed between them.

For **tall or heavy speakers**, larger sliders provide better stability and are easier to move across thick carpet. On hard surfaces, choose sliders with felt or polymer bottoms to prevent scratching.

Painter's Tape

Apply **wide painter's tape** to the floor for marking speaker and chair positions. In addition to tracking placements, it's helpful to place a strip next to each speaker to jot quick notes during setup.

Sharpie or Pen

Use **black Sharpie Fine** and **Super-Fine** tip pens. The **Fine** tip works for general notes, while the **Super Fine** tip allows for marking precise **dots** for laser alignment on the painter's tape.

Pliers or Strap Wrench

You'll need a **tool to grip and rotate the speaker spikes or feet**. Choose pliers with **plastic teeth** to avoid scratching finishes. For larger spikes, a **strap wrench** offers a more secure grip.

If your spike includes a small hole, insert an **Allen wrench** or **metal rod** for controlled turning. Precision here ensures stability and consistency in the speaker's stance.

Tape Measure

Use a **high-quality tape measure** to determine the **room's centerline** and **speaker distances**. For accuracy, avoid mixing brands or models—markings can vary slightly.

Recommendation: The **Klein Tools tape measure** offers durable construction and precise, consistent markings.

Steel Ruler

A **steel ruler** is ideal for measuring **speaker height.** Look for a ruler with a **sliding pocket clip**, which can be positioned as a **marker** for quick reference.

Scissor Lift Table

A small **4" x 4" scissor lift** can be invaluable. Use it as a **height gauge** for the front of the speakers or as an **adjustable platform** for a laser level. It's particularly handy for working on uneven floors or when your speakers have non-flat tops.

Bamboo Board

A simple **6" x 9.5" bamboo cutting board**—available from IKEA or similar stores—serves multiple purposes. Place it beneath a **laser level** on soft carpet, or use it as a **flat, stable surface** held under the bottom of speakers with curved tops when taking digital level readings.

Precision Digital Level

A **precision digital level** is indispensable. A model offering at least **two decimal places** of accuracy is essential; **three decimal places** are preferable for maximum precision. Be aware that some levels showing two decimal places are really only 1.5 decimal places as they are rounding to the nearest 0.05.

We recommend the **Digipas DWL 1500-XY**. This level includes **Bluetooth connectivity,** allowing remote readings from your listening position. This feature is necessary because standing next to the speaker—even a few feet away—can shift the reading. The slightest change in floor pressure can alter the speaker's angle, sometimes by a measurable amount.

Always rely on readings taken remotely from your **phone** or **tablet** at the **listening position** for the most accurate results.

Note: The Digipas consumes batteries quickly due to its Bluetooth circuitry. Keep extra **AA batteries** on hand. If you see the display flicker or you are getting very unlikely, inconsistent readings change the batteries. Allow the unit to warm up for at least **20 minutes** before calibration and **recalibrate** at the start of each setup session.

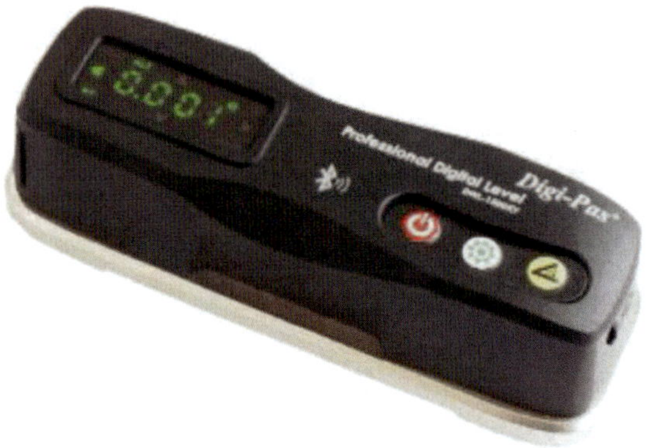

Self-Leveling Laser Level

A self-leveling laser level is essential for establishing the centerline of the room and assessing the height difference between the left and right speakers. Choose a model that projects both horizontal and vertical laser lines with narrow beam thickness for greater accuracy. Ideally, select one with the laser emitters offset from the main body.

A **remote control** is a valuable feature, enabling you to activate or deactivate the laser lines in any configuration without touching the unit, which helps preserve its precise position.

Safety First: The emitted laser can **damage eyesight** if viewed directly. Always wear protective glasses rated for your specific laser type. As an additional precaution, place a small piece of blue painter's tape over the upward-facing laser to prevent accidental exposure to your eyes.

There are many self-leveling laser levels on the market, each offering different degrees of precision. Generally, higher cost corresponds with greater accuracy and stability. Using this tool effectively requires patience and practice, but the payoff is substantial. When properly employed, it ensures that every speaker sits perfectly aligned within the room's geometry.

We consider the self-leveling laser level an indispensable tool for achieving true spatial symmetry and repeatable setup accuracy.

Laser Pointer/Level

A laser pointer/level is used along the side of the speaker to help assess and fine-tune toe-in with precision. The laser beam allows you to visualize the exact angle at which the tweeter is aimed relative to the listening position.

Low-Profile Jack or Inflatable Bladder

A low-profile jack is essential for raising and lowering the front or rear of your speakers when installing or adjusting spikes, footers, or sliders—especially when working alone.

One excellent option is the **LO-PRO** jack, available online. Pair it with a long-arm 7/16" ratcheting wrench to raise or lower the speaker smoothly and with control.

Safety Note: Always be mindful of your speaker's **tipping point** when using the jack. Although you should never raise a speaker high enough to risk toppling, carelessness at this stage can cause damage. Work slowly and deliberately, ensuring the speaker's center of gravity remains stable throughout the adjustment.

We recommend positioning the jack along the speaker's edge and using it primarily for small, precise adjustments, such as relieving tension on a spike that's difficult to rotate. Minor, controlled movements are all that's needed.

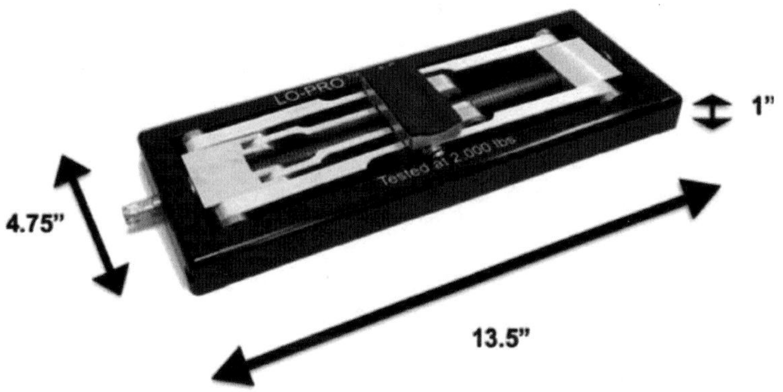

If the Jack will not fit under your speaker, use an inflatable bladder instead. These are slower but will get the job done. You can find these online or at your local hardware store.

With these tools assembled and ready, you can now begin establishing the initial placement grid, aligning your speakers, and performing the fine-tuning adjustments that will bring your system into full acoustic focus.

Summary: From Listening to Action

In Chapter 4, we recalibrated our expectations by examining how recordings are made and why that reality matters when evaluating sound at home. Once this is understood, many long-standing debates—particularly the false divide between accuracy and musicality—begin to dissolve. Accuracy gives music structure and coherence; musicality gives it emotion and meaning. Great reproduction requires both.

The reference tracks introduced here are not chosen because they are beautiful or impressive, but because they expose specific behaviors. These recordings form a practical listening vocabulary. They teach your ears what "right" and "wrong" actually sound like, long before you touch a speaker.

Most importantly, Chapter 4 shifts your role as a listener. You are no longer reacting emotionally and guessing why something feels off—you are observing, identifying, and diagnosing. That change is foundational.

With these listening skills in place, we are ready to move from theory to practice.

Section II begins the hands-on work of the **Precision Setup Method**. In the chapters ahead, you will learn how to translate what you hear into deliberate, repeatable adjustments of speaker position and listening geometry. Step by step, we will use the listening vocabulary developed in Section I to refine timing, spatial coherence, and tonal balance until the system begins to reveal the musical relationships embedded in the recording.

What follows is not guesswork or trial and error. It is a disciplined process of listening, adjustment, and verification—one that allows the stereo system to gradually disappear, leaving the music itself.

SECTION II

THE PRECISION SETUP METHOD

Why This Section Matters

Section II is where theory becomes action. Everything you learned in the first half of the book now converges into a clear, structured, and repeatable process for achieving **world-class loudspeaker performance** in any room, with any system.

This is not the typical "put your speakers 3 feet from the wall and toe them in slightly" advice. Those generic formulas cannot account for your room geometry, your speaker's dispersion pattern, your floor's rigidity, or your brain's sensitivity to micro-timing errors. Instead, this section teaches a precise method grounded in listening, physics, and thousands of real-world system setups.

The goal is simple: **to make your system sound like actual musicians performing in a real acoustic space.**

To reach that goal, you will learn:

- how to find the optimal listening position—the single most overlooked element of great sound
- how to establish reference points that ensure repeatability and symmetry
- why the right speaker must be trained first—and how its movements reorganize the entire soundstage
- how the left speaker becomes the reference instrument for tone, timing, and coherence
- how to integrate the right speaker so both channels behave as one unified source
- how to extend precision into the lowest octaves with subwoofers and bass towers
- how to treat the room in a way that enhances, rather than obscures, musical detail

By the end of this section, you will know exactly how to move a speaker in each of its six degrees of freedom—and, more importantly, how each movement *sounds* when it is correct or incorrect. Small adjustments will no longer feel mysterious. You will recognize the signature of misalignment as easily as a musician hears an out-of-tune instrument.

Once you hear what your speakers are truly capable of, there is no going back. Section II is your roadmap to that transformation.

CHAPTER 5

OBSERVATIONS REGARDING SPEAKER BEHAVIOR IN ROOMS

To place speakers accurately, you must first understand how they behave when they interact with a real room. Loudspeakers are complex instruments that project sound in multiple directions, across multiple drivers, and into a space that reflects, absorbs, and alters that sound in countless ways.

Chapter 5 introduces the fundamental behaviors you'll observe as you begin moving speakers through their six degrees of freedom: forward/backward, side-to-side, up/down, toe-in, rake, and azimuth. Each movement has a predictable effect on timing, tonal balance, coherence, and image placement.

This chapter sets the stage for all the hands-on work to come. By understanding how speakers respond in space—and how the room reacts in turn—you'll be better prepared to make precise corrections, interpret what you hear, and ultimately guide your system toward true musical coherence.

With these foundational concepts in place, we can now turn our attention to the practical steps needed to prepare both your room and your mindset for precise speaker alignment.

Preparing You and Your Room

A loudspeaker is a precision instrument, and like all such devices, it needs calibration. Calibration is necessary because, when loudspeakers are taken out of the box, they can produce sound, but they won't reproduce music accurately. For those who love turntables, this is similar to a cartridge. No one who has spent time with turntables believes that a cartridge doesn't need calibration.

Referring to our earlier discussion about the typical process of speaker placement, we observed that sound quality improves with more accurate positioning. I believe most people realize this early in their audiophile journey.

Granted, two issues prevent further progress: first, a lack of understanding of how precise we can be and the significant impact this precision has; second, the absence of a straightforward guide explaining how to proceed.

This chapter will likely be a radical departure from how you have previously viewed speaker placement. It will focus on the effects produced when the speaker is moved in each available direction. It also includes several observations about how the speaker behaves as it is continually pushed in one of those directions. These unusual phenomena are essential to understand if the goal is to find the best possible speaker orientation. Once these are explained, we will prepare everything to start moving the left speaker.

With the groundwork established, the next step is to understand how a speaker behaves when it actually moves through space—knowledge that will guide every adjustment that follows.

Moving the Speaker in Space

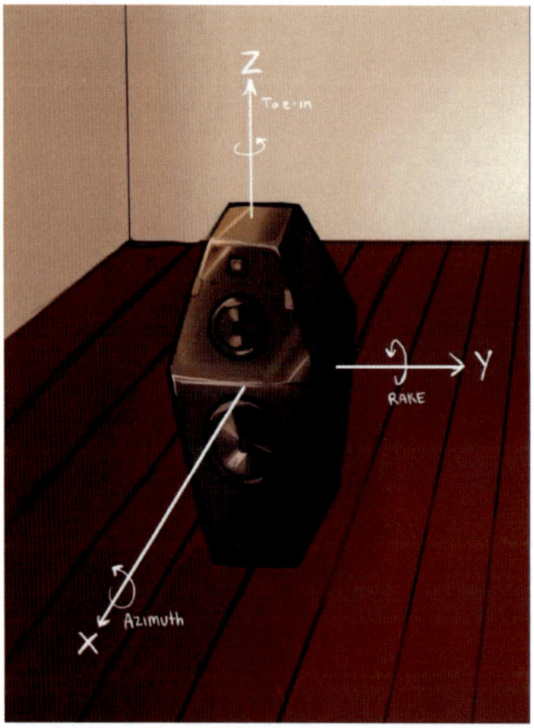

The speaker in your listening room is a three-dimensional object that can move freely within the room's three-dimensional space. Each independent direction the speaker can move is technically referred to as a "degree of freedom." All objects in three-dimensional space possess six degrees of freedom. The first three are translational: forward/backward, left/right, and up/down. Of these three, two are more commonly used, while the up/down movement is typically considered only during the initial leveling process. The next three are rotational: rotation around a line from floor to ceiling (toe-in), rotation around a line pointing from the side of the speaker (rake), and rotation around a line pointing through the front of the speaker (azimuth). Typically, only one of the three available rotations is utilized—toe-in. Thus, of the six degrees of freedom available, only half are usually employed. This leaves a significant amount of performance potential untapped.

To make these movements meaningful, it helps to distinguish between broad positional changes that establish the general behavior of the speaker and the fine adjustments that refine its performance. We will outline what to expect from changes to each axis. There are two categories of movement within each degree of freedom:

1. **Large movements**, where the speaker is being pushed or rotated to find an initial position.
2. **Small movements**, where the speaker is being bumped slightly, making corrections to refine the sound.

Degree 1: Front-to-Back — Pulling the speaker forward from the front wall

Large: Finding the correct position for the speaker in the front-to-back direction, away from the front wall, involves locating a position where the bass is locked in and the midrange sounds smooth and open. There aren't endless options available. Very few positions work, and most do not. Once in this position, the bass will not be boomy, and the midrange can sing. If someone were to master only the skill of finding the precise position where the speaker can coexist with the room, their sound would improve dramatically.

Small: If the treble is bright or too forward-sounding, bump the speaker forward. If the bass attack is slow, then bump the speaker forward. If the left-to-right speaker timing is off, bump the right speaker either forward or backward until it is correct.

Degree 2: Lateral — Moving the speaker side-to-side

Large: This is about time coherence and integration across a broad frequency range from bass to treble. When you find a good front-to-back location, the bass and midrange work very well in that position. However, this doesn't mean that the bass, midrange, and treble all work together in time with the room. The goal of finding a suitable lateral position is to align all frequencies broadly in time with one another. Lateral movements also contribute to a more "quiet" sound by lowering the acoustic noise floor. Aligning the lateral grid removes the opaqueness that obscures hall-space texture, allowing for clarity in male and female vocals.

Small: If the sound is legato, hazy, fuzzy, or "clashy," bump the speaker laterally (typically out). Pay attention to the hall sound—it is very delicate and can be obscured by a lateral movement.

Degree 3: Vertical — Adjusting the speaker's height (up/down)

Large: This involves enhancing bass articulation and refining bass-to-midrange timing. Begin with the speaker set at its lowest possible height while keeping it level. Turn all four (or three) spikes to raise the speaker in small increments—approximately 1/8 to 1/4 turn at a time. As the speaker rises, the bass will evolve from muddy and murky to more articulate. Raising the speaker will clear the bass's "ground fog." We aim to find a position that optimally balances bass weight and articulation.

Small: If the bass is stuck on the floor (ground bass fog) or inarticulate, raise the speaker. If the bass-to-midrange timing is off within a speaker, use vertical adjustment to correct the timing.

Degree 4: Toe-in — Rotating the speaker around the vertical axis

Large: Toe-in adds focus and achieves spectral balance among the bass, midrange, and treble. Finding a speaker's precise toe-in balances a singer's voice. If it's too much, the voice sounds nasal; if it's too little, the voice is too "chesty." Notably, there is a small window in the toe-in where tweeter and room noise are significantly reduced, allowing the voice to gain greater clarity. The sound takes on a "quietness" or "stillness."

Small: If the sound is too fast, noisy, or lacks "quietness," adjust toe-in slightly.

Degree 5: Rake — Rotating the speaker forward or backward

Large: Rake serves many functions. Finding the correct rake angle will control treble energy and release bass into the room. Discovering the small window where the rake angle is perfect significantly reduces acoustic noise. Raking a speaker also has a dramatic effect on image height. For example, if the tweeter is too low or the speaker is projecting sound downward, a singer can sound as if they are on their knees when the speaker is perfectly level. Raking the speaker back will raise the sonic image of the singer.

Small: If the sound field lacks perspective or the images are the wrong height, adjust the rake angle. Adjust rake to align the timing between the midrange and the tweeter.

Degree 6: Azimuth — Rotating the speaker sideways

Large: Modifying azimuth alters the relationship between the tweeter and midrange, as well as the distance to the sidewall. This adjustment primarily serves tweeter time alignment, refining tone and spatial positioning of sonic images. Azimuth adjustments are usually minimal. If a speaker's azimuth exceeds 0.05 degrees, something is likely very wrong with the setup.

Small: If the sound is leaning or angled, or if the very top treble is unclear while other bands are clear, adjust the azimuth.

If you make an adjustment to one of the degrees of freedom and get an unexpected result—with nothing sounding like it did before—check the speaker feet to ensure they are stable. A destabilized speaker will destroy the sound. One foot may need a little tension.

Now that we've examined how each degree of freedom behaves independently, we must address a harsh reality: none of these adjustments occur in isolation.

Understanding Compensating Moves

What is a compensating move? Simply stated, whenever we shift the speaker by any amount in one of the degrees of freedom, at least one of the other degrees of freedom is affected. The consequence of this is that any movement does not have an independent effect. This complicates the speaker alignment process. For instance, consider moving the speaker laterally inward by ¼". This adjustment can alter many relationships: effective toe-in, the distance of the speaker to the side wall, the reflection behavior of the speaker, the spacing between the speakers, and the distance from the speaker to your ear. That's a significant number of changes! If you determine that the lateral move achieved your desired outcome, you must revisit your checklist of accomplished goals to ensure they remain intact. Let's consider this in more concrete detail.

If the speaker moves laterally inward (Degree 2), then toe-in (Degree 4) will certainly need to be re-evaluated since the tweeter is no longer pointed where it was before the lateral move. If you determine the speaker needs more toe-out to compensate for the lateral move, the amount of turning, however small, will also change the speaker's front-to-back position (Degree 1).

Therefore, you need to ensure that the bass is still locked in. Additionally, this will slightly reduce the distance of the lateral move (Degree 2) you just made, as you are rotating the speaker away from your ear and toward the side wall. Does this now-reduced lateral move diminish the specific result you achieved before you toed out the speaker? Moving the speaker laterally also alters its distance to your ear, so you must assess the timing of the tweeter to the midrange driver.

Rotating the speaker away from you to correct the toe-in might mitigate the problem because it takes the speaker farther from you, but from a different angle. Try to note down what you did to achieve each goal so that if you need to repair it with a compensating move, you know where to start.

Okay, that was a lot of words for something simple. Let me restate and summarize: moving a speaker in any of its six degrees of freedom has consequences and impacts at least one of the other degrees of freedom. You will need to go back and correct them one by one. This is a process of finer and finer adjustments. The bumps, nudges, and taps get smaller and smaller until you finally arrive.

Problems With Unevenly Spaced Speaker Feet

A special case of compensating moves occurs when the speaker doesn't have evenly spaced feet. If your speaker has a square base and perfectly evenly spaced feet, then you're fortunate and can bypass this section. Notably, most speakers have unevenly spaced feet, so we need to consider the implications of the manufacturer's choice. That is, adjusting rake and azimuth is no longer independent.

Consider a speaker with four feet where the rear spikes are closer together than the front spikes. The outline of the spikes forms a trapezoid. Lowering both inner spikes by the same amount will tilt the speaker outward by some angle. Granted, since the rear spikes are closer together, the outward tilt also affects the rake angle. The speaker will tilt forward in addition to outward. We must use our level to restore the original value of the rake angle before the azimuth was adjusted. This allows for the independent assessment of the change in azimuth.

Speakers with only three feet cannot independently adjust rake and azimuth. Determining which angle is inadvertently affected depends on the orientation of the three feet. For instance, we might have two feet in the front and one in the back, or two feet on one side and one on the other. The simplest solution is to use the digital level to ensure that only the intended angular adjustment occurs.

Just as the geometry of a speaker's feet influences how adjustments behave, the physical arrangement of its drivers introduces another layer of complexity that must be understood.

Arrangement of Speaker Drivers

Adjusting the speaker's rake, azimuth, and toe-in (collectively known as attitude) will require consideration of the speaker's drivers' spatial relationship. Speakers come in various shapes, sizes, and configurations of drivers. The addition of more drivers to a speaker makes alignment increasingly challenging. The easiest type of speaker to align is a two-way design, followed by the straightforward three-way configuration. This book's illustrations feature a traditional three-way speaker due to its popularity among consumers. It typically has the woofer(s) at the bottom, the midrange in the middle, and the tweeter on top. Adjusting this speaker style generally follows a linear path. However, not everyone has this type of speaker, and even if you do, you need to understand the implications of having the tweeter as the top driver.

Figure 8: Speaker Driver Arrangement

If there's an alternative way to arrange the drivers, then a speaker designer somewhere believes that's a good idea. Study Figure 8, which shows just a sample of various speaker designs. They are arranged from the simplest on the left to the most complex on the right. Some designs have woofers at the bottom, the tweeter in the middle, and the midrange on top. There are also speakers with four drivers that place a second mid-bass driver at the top. Panel speakers usually feature horizontally arranged drivers, with a tweeter ribbon on one side, the midrange ribbon next to it, and the bass panel following. Furthermore, some speakers have multiple drivers that cover different frequency ranges positioned above and below the tweeter. All these variations imply that any given speaker will behave differently than other designs. The speaker setup path remains consistent, but whether the speaker is raked forward, toed-in, or azimuth-out will depend on the specific model's behavior.

Adjustments will become increasingly complex and non-linear as the speaker itself becomes more intricate. In a non-linear process, unexpected outcomes arise when a series of adjustments are made to correct a specific anomaly. In these scenarios, we must consider the arrangement of drivers and determine which frequency range or timing we aim to align. Additionally, what occurs in the relationship between the drivers and our ears when we make a height, toe-in, rake, or azimuth adjustment? We must recognize that as the drivers are positioned higher on the speaker, they will exhibit larger movements when either rake or azimuth is adjusted.

When listening and deciding how to adjust the speakers, think of the tweeter as a violin, the midrange as a cello, and the woofer as a bass. You could also consider the tweeter as a snare or hi-hat, the midrange as a tom, and the bass as the kick drum. The specific reference music used dictates which instruments represent each driver. Physically examine your speaker. Almost all speakers have the woofer at the bottom, so that one is likely pretty easy to find. Where is your midrange relative to the tweeter? Is it above or below? Do you have a fourth driver covering a separate frequency band? If so, you may consider adding a fourth instrument, such as a viola, to the mix. As you change the speaker's orientation, the relationship of those drivers to your ear will change. If the tweeter is above the midrange, raking the speaker back will delay the tweeter's timing. But if the midrange is above the tweeter, raking the speaker back will delay the midrange timing relative to the tweeter.

Figure 9: Driver Arrangement Interactions

To illustrate how driver arrangement interacts with timing and spatial alignment, let's look at the real-world scenario depicted in Figure 9. While setting up a pair of Avantgarde DuoGT speakers, I had the position of the left speaker very well dialed in. The tone part of the equation was excellent. Nothing was sticking out or poking me in the ear. Instruments had a natural sound, but there was still no foot tapping, head bobbing, or swaying to the music. What was missing were the final timing adjustments. The missed timing cues caused a lack of rhythm. The decay was incomplete, resulting in a loss of emotion in the music.

I started playing a song with piano and female vocals: Katharine McPhee's "She Used to Be Mine." (Once you gain some experience, you start to understand how to use the timing cues in music to adjust speakers. It doesn't have to be a song in the reference list to provide information. Almost all songs can provide guidance on how the speaker should be adjusted once you know what to listen for.) The song opens with a slowly played, repetitive chord. There is plenty of space to hear each piano note's full decay. The chord's repetitive nature creates an exceptional feel for the song's tempo. Following the fifth chord, Katharine starts to sing. Many of the song's words align with the piano's beat. If the speaker is misaligned, the song will lack rhythmic pace. The decay of the piano will be obscured by the words Katharine sings because they will begin in the middle of the decay instead of on the beat. The song will sound rushed. We need to fix this.

The piano chord being played is F# major, with frequencies of 185 Hz, 233 Hz, and 277 Hz. This chord falls within this speaker's crossover region, where the woofer hands off to the horn at 170 Hz. This means that the piano notes are reproduced from both the woofers and the midrange horn. Meanwhile, Katharine's voice is clearly situated in the upper horn's range. Listening to the song revealed that either the piano was significantly delayed, or Katharine was early. (Care must be taken here because these are humans playing music. The timing is not perfect; on many of the words, Katharine leads the piano just slightly. This gives the music a touch of swing.) At this point, I faced a decision. Should I delay the midrange timing, advance the woofer timing, or do some of both? To advance the woofer timing, I would slightly raise the speaker height (Degree 3) 1/32 or less of a thread turn at each of the four spikes. Alternatively, I could tilt the speaker back (Degree 5) by lowering the rear spikes just a little.

Or I could perform both actions at once by raising only the front spikes (Degree 3 + Degree 5). We must also remember that tilting the speaker back will affect the midrange-to-tweeter alignment. Since the midrange is positioned above the tweeter for this speaker, angling it backward will delay the midrange in relation to the tweeter. I used an EDM song with three evenly spaced midrange tones played like a triplet to obtain this timing cue. After several bars of the triplets, a very sharp, high-pitched "snap" is repeatedly heard. This snap occurs precisely at the leading edge of the second note of the triplet. Katharine's voice and the piano must align, AND the snap in the EDM song needs to line up as well. Therefore, we have a narrow needle to thread to correct both.

At first, listening to the EDM song, the "snap" made no sense. The timing was so poor that it seemed to occur randomly. Sometimes it happened before the first note of the triplet and sometimes after the second note. It felt random. This caused a great degree of uneasiness and anxiety while listening to the music. There was no way to get into "the groove" with it. I needed to bring Katharine and the piano closer and then worry about the "snap." Why Katharine first? This is because getting the lower frequencies right requires more significant movements. Then, the tweeter movement is smaller and will fit within the timing of the piano.

So, I raised (Degree 3) the speaker slightly and listened. I did this a few times, and I could clearly hear how Katharine's voice was more in sync with the piano's beat each time. I began to sway with the music and "feel" it. The song was starting to slow down, and the raw emotion was beginning to come out. After a couple more small adjustments, it sounded really nice. The decay seemed to last forever. I could truly lose myself in the song. I switched to the EDM track, and to my disappointment, the "snap" still wasn't lined up, but it was close to the second beat of the triplet. This only required a very small rake forward (Degree 5), and voila — music now makes sense.

This practical example sets the stage for one of the most surprising aspects of speaker behavior—a phenomenon that reveals itself only through careful, repeated observation.

Circular Timing and Alignment Zones

The observed phenomenon described in this section is likely speaker setup's strangest effect. I fully expect that when you read this, you will scratch your head. However, the described effect is fully repeatable and demonstrable. To state it succinctly, all the adjustments you make to the speaker (front-to-back, lateral, height, toe-in, rake, and azimuth) have multiple solutions that may seem correct. Think of them as multiples of the first solution. It is often the case that every other possible position along a path is incorrect, meaning, for instance, only the even-numbered slots along a path will actually work. There is not an infinite number of solutions. A few examples are needed for clarity.

As mentioned earlier, the lateral movement (Degree 2) of a speaker is responsible for aligning the treble, midrange, and bass at our ears. (PRAT is impossible in a system unless the speaker is laterally positioned correctly.) Using 10,000 Maniacs' "Like the Weather," we want the snare drum's snap to align with the bass line. Initially, this song sounds bad and almost unrecognizable. As the speaker is pushed sideways, the syncopated rhythm begins to develop. The continued lateral movement improves the rhythm until we finally arrive at a position where everything is in time and the music makes sense. The song will have much more snap and drive in this position. Interestingly, if we continue moving laterally, the timing and sense of rhythm will gradually degrade. If we keep pushing the speaker, we will find the position where the timing is at its worst, and the song will become unrecognizable once again. Keep pushing the speaker sideways along the path, and the timing and sense of rhythm will improve again. Eventually, we find another position where the timing is perfect, and the song has a strong sense of drive and rhythm. At this point, we have discovered two positions where the timing is correct. We can continue out (or inward) and find many such positions.

What is happening here? As the speaker moves laterally, it seems we are aligning the speaker's direct sound with the reflected sound from the side wall. This is basic grid alignment. (Recall the grid model presented earlier.) The room grid and the speaker grid align at the points where the bass, midrange, and treble are in sync. We experience cycling from good to bad and back to good again as the bass and treble move in and out of phase relative to one another while we push the speaker laterally. This phenomenon is observable whether the wall is untreated or has some acoustic panel at the reflection point. It becomes clear why every other point we perceive as "in time" does not work. While the treble and the bass are in sync with one another, the polarity is inverted, causing a strange phase shift to occur.

Here is a real-life example. I recently set up a pair of speakers. The speakers were already positioned very well, but I wanted more boogie. I moved them laterally (Degree 2) inward, searching for the ideal spot. This adjustment happened to be almost exactly ¼" away. I sat down and listened to "Like the Weather" and was stunned to find that it sounded like someone had turned the tweeter off. There was no organ to be heard. I stood up, scratched my head, continued along the path, and moved the speaker another ¼" laterally inward. Wow! Not only had the organ returned, but there was the boogie I was looking for. I share this story for two reasons. First, to give you an idea of how subtle the lateral distance can be. It is not always ¼", as that depends on your room and the distance from the wall the speaker is sitting. However, the distance is typically less than ½". Second, it illustrates the profound difference between good and bad positions, even though both have the correct timing.

A similar thing happens for the other movements.

- **For front-to-back movement (Degree 1)**, it is common for every other spot to be suitable for the bass. One will produce the sound of bass, while the other will provide the sound along with the grip or growl.
- **For height (Degree 3)**, you typically start low with heavy bass fog on the ground. The bass will sound very muddy and inarticulate. Raise the speaker by 1 to 3 mm, and the bass will become lean and articulate. Continue raising the speaker, and the bass will thicken and become congested again. Keep going, and it will lean out once more. Most speakers don't have the spike length to find the second correct zone.
- **For rake (Degree 5)**, you can identify zones where the soundstage is level from right to left. Continue raking the speaker, and one side will rise higher and higher, then suddenly drop and be slightly lower. You can keep raking the speaker until the soundstage is level once more.
- **For azimuth (Degree 6)**, the treble timing will move in and out, but as you continue to adjust azimuth in one direction, the treble and midrange will overlap. This is the grid concept applied to azimuth. Be cautious with too much azimuth. It should be used sparingly to fix very specific problems at the end of the setup.

It is easy to find yourself in a false zone (grid) within one of the degrees of freedom. A false zone occurs when that degree of freedom sounds correct, but in reality, the speaker is one or two grids off. If you are working with the speaker and feel stuck, and it seems impossible to make progress or things aren't moving linearly, then ask yourself if you believe the speaker is in a false zone. Identifying which degree of freedom is in a false zone can be challenging, or by this point, you may have multiple degrees of freedom that are off. You need to listen to various songs to determine which one is incorrect. Being in a false zone is one reason the right speaker must be positioned as accurately as possible in the "start" position. If it is placed too far off, you might move it into a false zone, assuming it is correct when, in reality, you're very far off. We will discuss this further when we determine the right speaker's position.

With these pitfalls in mind, we can now bring together the individual concepts and explore how they interact to form a coherent framework for speaker alignment.

Combining Ideas

We can combine the ideas of each speaker driver—representing a specific instrument—with the concept of alignment zones. Think of this relationship like a Chinese cylinder puzzle, where multiple rotating segments must align perfectly to solve it. In our acoustic version, the tweeter represents one cylinder, the woofer another, and each has a corresponding "room interaction" cylinder that represents how that driver's output behaves in your specific listening space. For optimal coherence, all four elements must be aligned—or, in acoustic terms, they must all be "in phase" with proper timing relationships.

When drivers fall out of alignment, the most problematic situation occurs when they are 180 degrees out of phase—essentially working against each other rather than together. This is like having one cylinder facing completely backward relative to the others. Granted, as you work to correct this misalignment, you'll discover that there are multiple positions where the drivers can achieve proper timing. Just as you can rotate a puzzle cylinder through complete revolutions and find several correct alignment points, acoustic phase relationships can be adjusted through full 360-degree cycles while maintaining coherence.

The crucial insight is that while these different alignment positions may all be technically "correct" in terms of timing, each produces subtly different tonal characteristics. The interaction between drivers and room acoustics shifts slightly with each complete phase cycle, affecting the overall timbre of the system. This means that achieving perfect speaker coherence isn't just about finding any correct alignment—it's about finding the specific alignment that produces the most natural and pleasing tonal balance for your particular room and listening preferences.

Sound Direction

Loudspeakers project sound forward toward the listener. This is obvious to most people. Most notably, what isn't so clear is that the sound has a vertical direction. Once the sound from the different drivers is a few feet from the speaker, it is merged into a semi-coherent wavefront. Depending on the speaker's design, the wavefront can travel upward or downward. When a typical speaker is placed and leveled in a room, the wavefront usually travels downward, and the sound hits you at the knees. We need to raise it so that it hits the upper torso and head. This is achieved by adjusting the rake angle. (It could be done by height, but most speakers do not have the spike length needed to achieve this. The two methods will produce different results since the sound is still projected downward if the height is raised.)

I believe the cause of this is related to crossover design and the extent of overlap between drivers. In the three-way design shown below, any crossover present will cause the tweeter to overlap with the midrange and the midrange to overlap with the woofer. When the midrange plays part of the tweeter's range, it can influence the sound in that direction. The same applies to the woofer. For example, there is a well-known speaker from the 1990s that used first-order crossovers. This speaker was very tall, with the tweeter positioned well above seated ear level, yet the sound was projected at your feet.

With a D'Appolito (MTM) design, we have the drivers symmetrically placed above and below the tweeter. In this case, there are two lobes. One is projected upward and the other downward. If you are working with an MTM design, you get to choose which way the speaker is raked (forward or backward). Each choice will produce different sonic results. You should experiment with each to see which you prefer.

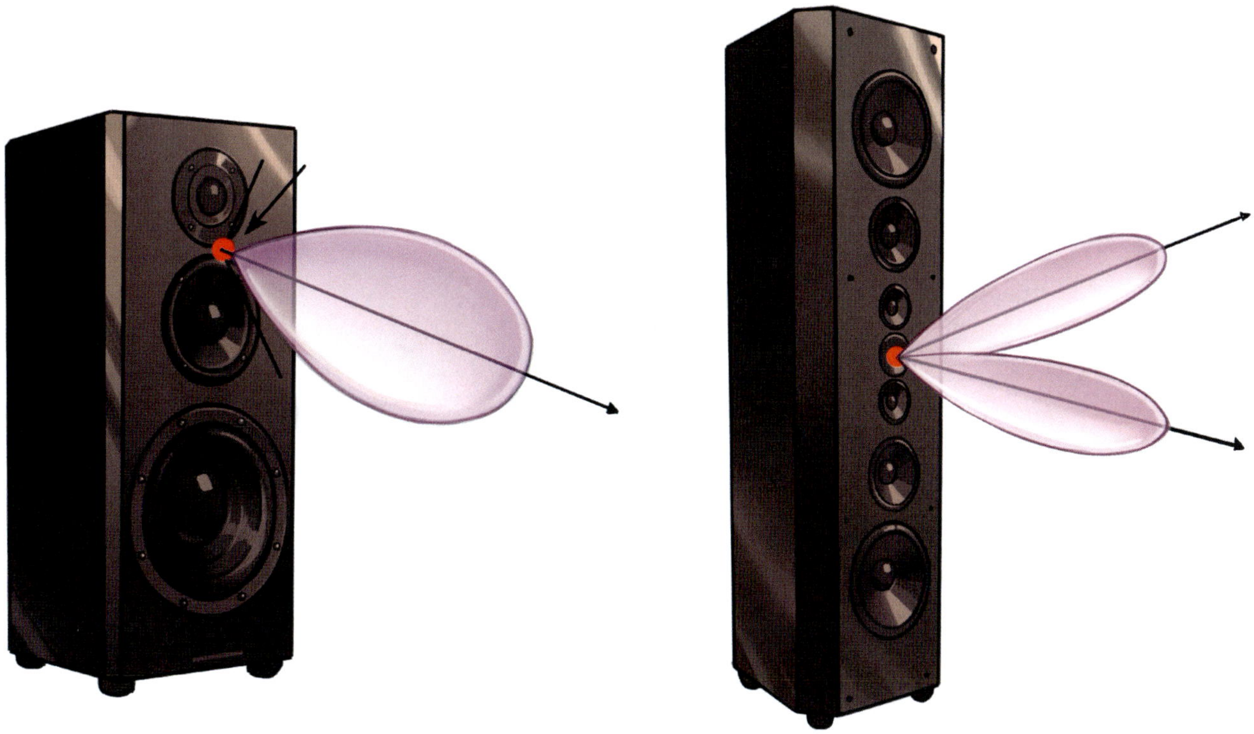

Summary: The Loudspeaker as an Instrument

Chapter 5 established a central truth that underlies everything that follows: a loudspeaker is not simply a device that produces sound. It is an instrument whose performance depends on precise orientation in space.

Throughout this chapter, we examined how speakers interact with the room and why their physical behavior cannot be separated from the environment in which they operate. Small changes in placement, angle, height, or distance alter the timing relationships between drivers, the interaction with room boundaries, and the way sound waves combine before reaching the listener. These changes can dramatically affect tonal balance, spatial stability, and the overall coherence of the musical presentation.

Understanding this transforms how we think about speaker setup. Placement is not a cosmetic adjustment or a matter of convenience; it is the primary means by which a playback system reveals or obscures the information embedded in a recording.

With this principle established, we can now turn to the other half of the equation: the space in which we listen. Even the most carefully aligned loudspeaker cannot reveal its full potential if the listener is seated in the wrong place. The room—and our position within it—determines how sound waves interact, how bass pressure develops, and how the stereo image forms between the speakers.

For this reason, system setup does not begin with moving speakers. It begins with understanding the listening position and the acoustic geometry of the room.

Chapter 6 begins the practical journey by addressing the most fundamental question in system setup: **Where should the listener sit, and where should the speakers begin?**

From this starting point, we can begin the systematic process that will guide the rest of the Precision Setup Method.

CHAPTER 6

THE LISTENING POSITION AND INITIAL SPEAKER POSITION

With the noise removed and the music finally revealed, we now turn our attention to the space in which we experience it. The room itself—and where we sit within it—plays an equally critical role in how we perceive tone, timing, and dynamics. Even a perfectly optimized system can lose its magic if the listening position isn't right.

Just as a photographer frames a scene to balance light and shadow, we must find the location in the room where sound waves converge in harmony—where bass is smooth, imaging is stable, and the performance feels lifelike. That leads us to the most fundamental question of all: before we move the speakers, where should we sit to hear the truth of the recording? This chapter shows you how to find that spot.

We'll begin by identifying the room's natural acoustic behavior—where bass gathers, where it cancels, and where it becomes most balanced. From there, we'll mark the true centerline, establish a repeatable geometry, and position your seating so that the listening environment is primed for precision work.

Only once the listening position is correct can the speakers be placed with confidence. In the chapters that follow, every movement—every bump, nudge, and micro-adjustment—will reference the foundation you establish here.

Finding the right seat isn't just step one. It's the anchor for the entire Precision Setup Method. Let's begin there.

Optimizing the Listening Position

In some rooms, the listening position may be fixed to a single location. If that's the case, proceed directly to optimizing speaker placement. Although, if space allows, the position of the sofa or chair can—and should—also be optimized. A large adjustment is rarely necessary; even a small movement of a few inches or one to two feet can make a meaningful difference. Whatever flexibility your room provides should be viewed as an opportunity for improvement.

Jim Smith discusses this concept in his excellent book *Get Better Sound*. If you have not read it, I highly encourage you to purchase a copy. Below is a brief summary of the process he describes, along with additional insight.

Tools and Setup

To put these ideas into practice, we need a reliable method for identifying where the room naturally supports smooth, stable bass—and that begins with a few essential tools: laptop, microphone, cables, and acoustic measurement software. *Room EQ Wizard (REW)* is free software designed for recording and processing acoustic data. It includes both a pink noise generator and a real-time analyzer (RTA), and there are many helpful instructional videos available online.

If *REW* feels intimidating at first, several smartphone apps also include RTA functionality. Pink noise can be streamed from many sources or found on numerous setup CDs. However, I still strongly recommend using a dedicated measurement microphone rather than the one built into your phone or tablet. While smartphone microphones are acceptable for a quick reference, they lack the accuracy needed for precise setup.

The Basic Method

The process is simple yet powerful: play pink noise through your speakers and use the RTA to evaluate bass smoothness. Move the microphone along the room's centerline until you find the position where the frequency response is smoothest. With the purpose of the test understood, we can now walk through the process in a clear, structured way. The remainder of the procedure is simply the "how" of performing each step.

Setting Up the Software

Set the RTA to **1/3-octave resolution**. Higher resolutions can appear confusing. Once you have identified a promising position, you can increase the resolution to **1/6-octave** for fine-tuning.

Focus on the frequency range between **20 Hz and 150 Hz**, since this is where room acoustics most strongly affect the sound. In *REW*, set the frequency window accordingly.

Start playback of the pink noise and press *Record* to begin analyzing. The RTA bars will fluctuate continuously. In theory, a perfectly smooth frequency response would form a straight line—but this is never the case in real rooms. The goal is to find the flattest, most consistent response possible.

With this visual framework in mind, the next step is to begin taking measurements that reveal how your room actually behaves.

Finding the Position

Once setup is complete, play pink noise through your speakers and hold the microphone at seated ear height. Move slowly along the centerline of the room. Begin with larger movements—about **6 to 12 inches**—to understand the general trend. The RTA bars will rise and fall as you move.

As you progress, narrow your adjustments to within an inch or two, searching for the position where the bass is smoothest (not necessarily deepest). The optimal location will show minimal fluctuation, without large peaks or dips.

If you must choose between a significant bump or a dip, **choose the dip**. Your brain can easily ignore a small reduction in bass energy, but an exaggerated peak will dominate perception, masking other musical details and throwing off tonal balance.

Take your time—there's no need to rush. You can fine-tune this to within an inch for maximum precision. Once you locate the smoothest position, mark it on the floor with tape. Move your chair or sofa so that your **ears are directly above** that mark when seated, not the front or rear edge of the seat.

Once you've identified the smoothest position and marked it, the measurements must now be interpreted to confirm that the listening position is truly optimized.

Assessing the Result

Now it's time to evaluate what you've achieved. Play several familiar tracks, ideally ones with rich, natural bass. You should notice smoother, more linear bass reproduction. You may not hear deeper extension—that wasn't the goal—but the bass should sound more even and integrated with the midrange.

If you're satisfied with this position, move on to optimizing your speaker placement.

To illustrate the dramatic impact of listening position on bass, consider the Figure 10 example on the following page. The **blue line** represents the frequency response (up to 110 Hz) when the speakers are positioned at the front of the room, and the listening position is near the center. The **black line** shows the response after moving both the listening position and speakers approximately ten feet backward—placing the sofa about **five feet** from the rear wall and the speakers about **eleven feet** from the couch.

Notice how smooth the bass becomes, extending linearly down to **40 Hz**. The improvement in tonal balance and bass integration is substantial. The difference in musical satisfaction is equally striking—the bass feels natural, powerful, and engaging.

For bass extension below 40 Hz, adding a **subwoofer** can provide the additional low-frequency foundation needed for full-range reproduction.

Figure 10: Frequency Plot Illustrating the Impact of Moving the Listening Position (1/12 Octave Smoothing)

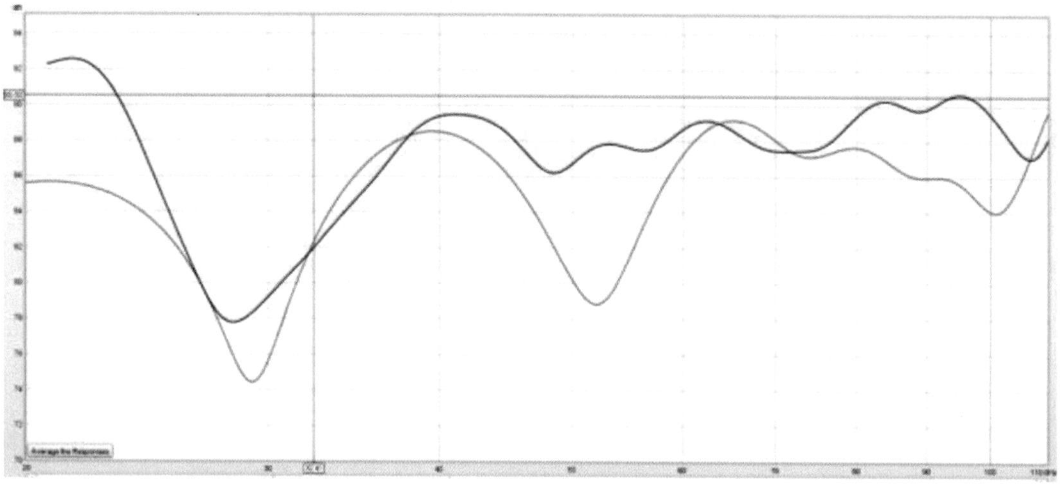

Moving the listening position **ten feet** back is not always necessary to achieve smoother bass. In many cases, a much smaller adjustment—between **8 and 16 inches**—can yield a significant improvement. Don't be discouraged if your room offers limited space to experiment. Even modest shifts in seating position can make a measurable and audible difference. Explore the available range in your room and see what gains can be realized.

With the listening seat anchored in the room's most supportive location, we can now establish the geometric framework the speakers must align to.

Setting Up the Starting Position

The next step is to find the room's centerline or the centerline of the area where you will place your speakers. You will sit on this line. If the room is symmetrical and you position the speakers in the middle, use a tape measure to find the center by measuring from each wall. Make sure to measure at floor level, since walls are often not perfectly plumb. Be aware of where the tape measure's end touches the wall; errors can happen if there are baseboards or uneven surfaces. It's important to get this measurement as accurate as possible. If your room is asymmetrical, measure from one wall to the center of the listening chair. Use that measurement to find the centerline. Place a piece of blue tape on the floor, centering the tape on the approximate center. Then, press the tape measure firmly against the wall and with a Sharpie super-fine pen make a small dot on the blue tape where the room or system's center will be. Move the tape measure approximately 12 inches and repeat for the second dot, trying to apply consistent pressure and be as precise as possible. These two dots should accurately represent the center. Variations in baseboard thicknesses or wavy walls may occur but using the side wall as a reference point is a good choice.

Note: Be aware if you place the blue tape on carpeting or an area rug that the rug (or the nap of the rug) may not stay in place. If you must place the blue tape on a rug, then try to be consistent in how the tape is positioned (left or right) when you mark and read the markings.

If you mark two points that are, for example, 78 ½" from the side wall baseboard to the system center and 78 ½" to the center of the listening chair, and a laser lines up with all three dots, it's a good day! If they do not line up, place a few more marks at 78 ½" and select the ones that align. Just remember that the two dots on the floor you choose must also align with the center of the chair. When the three dots align, that becomes your centerline. Aligning the speakers to the centerline will position them symmetrically relative to your seated position, which is essential for the setup process. We have encountered many instances where the center of the listening chair is three or even six inches off to one side of the centerline, which we want to avoid. The setup process relies on the symmetry of the speakers to achieve the best results.

With the room's reference line defined, we can begin positioning the left speaker—the foundation upon which all further alignment depends.

Initial Left Speaker Position

Determining where to start with the speakers can be confusing. The speaker can sit in countless positions within a room. There are plenty of suggestions, including the "golden ratio," "Rule of Thirds," "Zone of Neutrality," and more. Granted, none of these methods will pinpoint the exact spot where the speakers perform best in your room. Finding the ideal position for your speakers is what we will explore. Due to domestic constraints, most speakers will be placed within a few feet of the front wall. I recommend using a piece of painter's tape to mark the floor, indicating the farthest position from the front wall that you (or someone else) are willing to accept for the speakers. This will reveal the available space to optimize the speaker's placement. It's a good idea to move the speakers to that farthest position to ensure everyone is satisfied if they end up there.

As you sit closer to the speakers, you will hear more of the speaker's direct sound and less of the room. A good listening distance is typically between 9' and 12' (3 to 4 meters) from the speaker. Therefore, a good starting distance between the speakers is 8' to 10'. If the speakers are too far apart in relation to your listening distance, they will likely not couple together laterally. Based on hundreds of speaker setups, if the distance between the speakers exceeds about 85% of the listening distance, there is a high likelihood of improper coupling.

To find the starting point, some math is necessary. Measure the horizontal distance from your ear to the blue tape placed on the floor (perpendicular to the front wall), which indicates the speaker's farthest position from the front wall. Take that distance and **multiply it by 0.47**. This gives you the distance from the center line to the starting lateral position of the **left speaker**. For example, if your ear is 9 feet from the painter's tape, the starting lateral position would be 9 x 0.47 = 4.23 feet. When the speaker is moved further behind the painter's tape, the distance between the speakers will represent a smaller percentage of the listening distance. This will allow for some flexibility in the lateral position. (Of course, as you become more familiar with you room and the setup methodology you can experiment with a wider speaker placement.)

Using the number calculated above, measure from the center line of the room towards the left speaker. Place a long (18"-24") piece of painter's tape on the floor, perpendicular to the front wall. A good starting position is about one to two feet from the front wall. Mark the painter's tape to indicate the position of the center of the front inside spike. Write an "S" next to the mark. This will indicate your starting position.

Position the centerline of the left speaker's front inside foot at the starting position marked on the tape. Leave a 1/8" gap between the tape and the edge of the slider. This allows for lateral position adjustment without the speaker resting on the tape. Toe in the speaker to the "correct" angle (Covered in setting the left speaker). Adjust the spikes or feet so that the speaker is level both front-to-back and side-to-side, ensuring that the speaker is as low as possible.

Position the **right speaker** several inches back and outside of the 4.5' mark. (Covered in the initial right speaker section) Angle the right speaker appropriately based on its design type. Ensure the right speaker is level; we don't want it to wobble. Even when angled away, sound from the right speaker is necessary for determining the optimal position for the left speaker. (See illustration)

Once the left speaker is roughly established, the right speaker must be placed in a very specific temporary orientation that allows the left channel to be judged accurately.

Initial Right Speaker Position

You might be wondering about the right speaker—why it's turned to the side and what role it plays at this stage of the process. That curiosity is entirely natural; honestly, it does look quite odd. Positioning the right speaker pushed and turned away while working exclusively with the left is perhaps the most distinctive aspect of the speaker placement method outlined in this guide. In fact, it's a defining feature of the entire process. The reasoning behind it might not make complete sense yet, but as you work through the optimization steps, it will become clearer, just as it did for me. This method has been successfully applied thousands of times in a wide variety of systems, so for now, trust the process and follow along.

To clarify, the right speaker continues to play music along with the left throughout the process. Nothing changes in its audio signal; it's simply rotated away and pushed slightly out of position. This setup allows it to generate broad sound pressure across the right side of the room. Even when turned away, the right speaker contributes valuable information about how the left speaker interacts with the space—particularly on the right side. This arrangement helps us determine when the left speaker is properly aligned and integrated. If the right speaker were pointed directly at the listener, it could obscure or distort these judgments. Turning it away ensures that any problems we hear result from the left speaker alone, not its interaction with the right.

As the left speaker becomes more coherent and integrates better with the room, the soundstage will expand and fill the space between the speakers. Once this tuning is complete, the imaging and spatial realism can be astonishing—even with the right speaker still turned away. It seems counterintuitive, but the results speak for themselves.

Here are a few practical notes on positioning the right speaker during this phase. It should be pushed back slightly from the plane of the left speaker and angled outward. The extent to which the right speaker is rotated depends on its design type. The reason is that different designs (bass reflex, horn, panel) have varying amounts of off-axis response. We need to hear some of the treble information from the right speaker, so the degree to which we rotate it will depend on its design.

- Acoustic suspension or bass reflex speakers require a lot of rotation. A good reference point is to rotate them until you can barely see the midrange driver from your listening position.
- Horn speakers will need much less rotation. Adjust them so that you can see a good distance down the throat of the tweeter horn but cannot see the tweeter.
- Panel speakers are the most difficult to prescribe. Since the sound waves are radiated fully backward as well as forward, the speaker is strongly interacting with the room as it is turned away. You are looking for a position where the right side sounds very diffuse. Don't be surprised if you are looking at the panel's edge.

Interestingly, the precise location of the right speaker isn't critical at this stage—its job is simply to stay close to its final position without interfering. A placement within one to three feet of its final spot is usually sufficient.

Check the Cables

The last step is to ensure the cables are connected to the correct channel and in the proper polarity. We have all accidentally crossed channels or polarity at some point. With this setup method, it is difficult to determine if either is correct during the first phase of the setup. This is critically important: The various queues described in the setup process will not be present if the channels are reversed or the speakers have incorrect polarity connections. You would have nothing to listen for.

NOTE: *Correct Absolute Polarity means that the + and - connections at the amplifier connect to the + and - connections on the speakers. Out of Absolute Polarity means BOTH speakers' + and — connections are reversed from the amplifier's + and — connections. Out of Relative Polarity is where only one speaker has the + and — connections reversed.*

There are several ways to test for channel and polarity correctness. For Left/Right channel correctness, you can play Yello's "Till Tomorrow." The trumpet should start on the right and move to the left. Alternatively, you can use "Cyril Davies," where the saxophone is on the right and the trumpet is on the left. Any vocal track, such as Ruth Moody, can help check the phase. She should sound centered between the speakers. If she sounds diffuse and "stretched" from left to right, you likely have one of the speaker cable's black/red connections swapped. Checking for absolute polarity can be trickier to determine by ear, so the best way to ensure its correct is to look at the amplifier and speaker connections and verify they are where they should be. Yeah, we know that's a "Thank You, Captain Obvious" thing to say, but we still have to say it... In either case, turning off the amp is best when making speaker cable connection changes. You can mute it, but it's safer to turn it off.

Moving The Speaker

Optimizing speaker placement obviously requires moving the speakers. To do this with spikes or feet on the speaker, we must place them on sliders. You can create some using furniture sliders purchased at your local home improvement store, or you can buy a set online designed specifically for speaker spikes. Herbie's Audio Lab is a company that sells some nice ones. The sliders need a wider base if the speakers are tall and on a carpeted floor, as we don't want the speaker to fall over when pulling it forward. Regardless of which sliders are purchased, they must be positioned under the speakers before we can start moving them around.

The ease with which the speaker moves depends on its weight and the friction between it and the floor. Each speaker and floor is slightly different. We want to push, pull, or tap the speaker in a controlled manner as best as we can. Getting a feel for moving your speaker on your flooring will require time and experimentation. Any effort to move the speaker is best achieved at or near the lowest part of the speaker cabinet. This way, you can precisely position the speaker wherever or however you desire.

Recalibration of what constitutes a large movement is necessary. Our definition of a large movement is that if we can see the speaker physically move, then it qualifies as a large movement. We will begin with large movements. Initially, we will have movements ranging from 1/2" to 1" when finding the correct position from the front wall. Then we will make adjustments of 1/8" to 1/4" to determine the correct lateral position. There will be some significant changes to the toe-in angle at first. Eventually, the "movements" will become so small that we won't be able to see the speaker move. These "moves" consist of small bumps with the fist, taps with the knuckle, or a slight twist to the spikes.

Throughout this book, I use the term "bump the speaker." Unfortunately, the English language lacks a more elegant word for this. The word bump seems to imply some accidental contact, as in, oops, I bumped into the coffee table. In our case, nothing could be further from the truth. We are attempting to make very controlled contact with the speaker in a specific location to achieve a predictable result. By bump, I mean to strike the speaker lightly. Sometimes, this uses the side of the fisted hand; other times, it might be with the knock of the knuckles. It could even be a very light tap with the index finger. These "bumps" are intended to nudge the speaker a very small, visually imperceptible amount in a given direction. The firmness of the strike for each bump, knock, or tap will depend on the speaker's weight and the floor on which it sits. For example, a 700-pound speaker on spikes through the carpet into an OSB subfloor will require significantly more force to move than a 100-pound speaker with smooth-bottomed, titanium spike shoes resting on a smooth marble floor. Part of the process is understanding how hard a bump, knock, or tap is required to nudge your speaker a very small but clearly audible distance. From experience, you may think you did not hit the speaker hard enough when, in reality, you hit it too hard.

I know what you are thinking because I once thought it too—"There is no way simply bumping a speaker can change the sound." It does. When we position the speakers with greater accuracy, very small changes have a significant effect. We can't hear these changes when the speakers are poorly placed and not optimized. This process is analogous to focusing a camera lens. For example, suppose a professional photographer has a camera lens with adjustable focus and is taking a soon-to-be bride's portrait. Initially, the image is extremely blurry, and the photographer twists the focus ring a small amount. The image remains very blurry, and the small focus adjustment isn't even noticeable. However, if the photographer makes a large twist to the lens and brings the image into closer focus, he/she can then make that same small adjustment and BAM! The image is now clear and sharp. The resulting sharpness and clarity depend greatly on the quality of the camera, lens, and the photographer's skill level. This is precisely what we are doing with the speakers. We are moving them more and more into focus, and when the speakers become focused, even tiny movements have a noticeable effect on what we can audibly perceive. Just like the quality of the camera and lens, the system's noise floor ultimately dictates how finely we can focus our audio camera.

Take a moment to explore your speaker. Where is a good place to put your hands to pull it forward? Sometimes, we can pull on the front or spikes to move the speaker. Other times, we can grip the sides or along the bottom edge. Wherever this may be for your speaker, find a comfortable way to position your body so you have precise control over how much you move it. Start by getting a sense of how easily it slides on your specific floor. The ease with which the speaker moves depends on its weight and the amount of friction between it and the floor. A very heavy speaker that sinks into a carpet and pad is quite challenging to move in small increments. Conversely, one of the easiest and most precise situations we've ever encountered was a 750 lb. speaker on a marble floor. It may seem that a lighter speaker, something under 100 pounds, would be advantageous since it's easier to position. It's exactly the opposite. The distances we need to move the speaker are very, very small. The mass of a large, heavy speaker prevents it from moving too far, which, as you'll soon discover, is all too easy to do. If you have a light speaker, you need to learn how hard to pull or strike it to make very minor adjustments. The lighter the speaker, the gentler the taps needed to move it into the final position.

Situational Awareness

We must be hyper-aware of everything around us when moving near the speaker. It is often the case that we feel a bit cramped. We have the audio rack, equipment, side wall, front wall, and cables. Be in a calm headspace when you approach the speaker. Move with intent. The last thing we want is to accidentally bump into the speaker with our elbow or knee while getting up or down. I suggest removing your shoes so you can be more aware of your feet. We also want to be mindful of any tools we are using, as they can scratch the floor, the speaker, or the equipment. Take your time, and do not rush.

The Music Playlist

Create a playlist using the listed reference songs. Listen to each song and determine its appropriate volume level. Some songs are recorded at a low level, so they need to be played at a higher volume than others. You should aim for a higher volume setting than you usually prefer. This will energize the room and make it easier to hear "noise" and errors in the sound. Your system will sound even better once you optimize everything and play music at your normal volume level.

It is important that every time you play an individual song, it is at the same volume; otherwise, you might think it sounds more dynamic (for example) simply because the volume is one notch higher. Note which songs have inverted polarity. If your DAC or pre-amp has polarity inversion, use it for those songs. I suggest taking a piece of paper and writing down the title of each song, the volume level, and whether it is inverted. This way, you will have a quick reference handy.

NOTE: Some pre-amps do not have an exact numeric reading for the volume. Instead, they have a simple rotary knob without any indents or markings on the faceplate. If your pre-amp is like this, you can use your Sharpie to make a mark on a piece of blue tape, tear it off, and place it where the volume setting is for that song. You might give each of these a numeric marking—1,2,3, etc. I find that for the reference songs, there are typically 3 volume settings, so it should not be too cluttered around the volume knob.

Before these physical adjustments can yield meaningful results, your ears must be trained to recognize the cues that indicate proper alignment.

Preparing Your Ear

We will proceed with the right speaker turned away and listening primarily to the sound from the left speaker. Before finding the perfect position for the left speaker, I encourage you to start by turning the right speaker and simply listening to the left speaker as it currently rests. (The right speaker is still playing music. It has just turned away.) Listen to all the various parts of the reference songs, make observations, and take notes on what you hear. Listen to some of your favorite music through the left speaker. What you are trying to do is become accustomed to the sound of the left speaker, as this may initially feel unfamiliar.

With your listening sensitivity awakened, there is one final consideration that must be addressed before the setup process begins in earnest.

Last Chance

At this point, we have everything ready to start the search for the perfect speaker location. If you haven't read the section on noise reduction, I encourage you to do so and take steps to minimize your system's noise floor. You can still optimize the speakers, but if there is noise in the system, it will be difficult to determine whether it comes from the speaker/room or the electronics. Ultimately, we will adjust the speakers to accommodate the noise, as it will be part of what we hear. Although, this will always involve a compromise, and the sound will not be as transparent and dynamic as it could be. Let me illustrate with a real-world example.

Recently I was setting up a pair of speakers using a streamer as the primary music source. I optimized the left speaker to the best that system would allow. I then plugged in several grounding and network filtering devices targeting the streamer. The impact was transformative. Prior to the devices the piano in Beethoven's first piano concerto appeared approximately at the tweeter level. Once the noise reducing devices were installed the perceived piano image dropped in space to just below the midrange driver. This would be about 8-12 inches. The piano gained a tremendous amount of body. This allowed me to hear further into the recording and make substantial improvements to the speaker's alignment. If I had proceeded without the noise reduction the end result would not have been as good.

Summary: Establishing the Listening Reference

In Chapter 6, we established the foundation upon which the entire Precision Setup Method rests: the listening position. Before speakers can be placed with precision, the room itself must be understood, and the listener must be seated where the room naturally supports smooth bass response, stable imaging, and coherent timing.

By identifying the room's centerline and locating the listening seat at an acoustically favorable position, we created a stable geometric and acoustic reference point. This position serves as the anchor for the entire setup process. Every subsequent adjustment—speaker distance, toe-in, rake, and integration—will be evaluated relative to this fixed listening location.

This step is critical because stereo reproduction is fundamentally a **relationship between the loudspeakers and the listener**. Without a defined listening reference, speaker placement becomes guesswork, and the timing relationships required for believable imaging and tonal balance cannot be reliably established.

With the listening position now set and the room geometry clearly defined, we are ready to begin the hands-on process of speaker alignment.

The next step is to begin training with a single loudspeaker—specifically the right speaker. By working with one speaker at a time, we remove the complexities of stereo interaction and learn to recognize the auditory cues that reveal proper timing, balance, and spatial stability.

Chapter 7 begins that training process.

d
d

CHAPTER 7

RIGHT SPEAKER TRAINING AND EXERCISES

Whether you've arrived here by following the book step-by-step or you've eagerly skipped ahead, this chapter marks the true beginning of speaker placement. Here, we focus exclusively on the right speaker—how its position affects what you hear, how it shapes the soundstage, and why it plays an outsized role in unlocking dramatic improvements in your system's imaging and timing.

If you want fast, noticeable gains, you're in the right place. Correctly positioning the right speaker can transform your listening experience, even if the left speaker remains untouched for now.

PART 1 — Why the Right Speaker Matters

Spatial Organization

Think of the right speaker as a tool for organizing sound in space. In a simple studio vocal track, the singer should appear dead-center between the two speakers. In a full orchestra—80 to 100 musicians precisely placed—you rely on the correct speaker position to place the musicians accurately across the soundstage, bringing realism and depth to the performance. Notably, spatial placement is only part of the story.

Timing Organization

Timing matters, too. Music is much more about when the sound arrives than where the sound comes from. Timing coherence determines whether an orchestra sounds unified or disjointed. To sound cohesive, the musicians must not only appear in the correct positions but also play together. When the right speaker is correctly aligned, it helps us achieve both spatial precision and temporal coherence. The right speaker's position heavily influences this temporal alignment.

The Phantom Center Image

Unfortunately, a common problem in many systems is that the center image drifts slightly left or right. This is almost always the result of improper left/right symmetry—and it is surprisingly easy to correct once the right speaker is properly positioned. It simply requires the speakers be placed symmetrically about the centerline of your listening position.

When the left and right speakers are not correctly aligned relative to one another, the result is frequency-dependent imaging. For example, while listening to a symphony, the clarinet might drift—sometimes appearing near the horns, other times near the violins—depending on the pitch of the note. This inconsistency is incorrect; an instrument's location should remain relatively stable.

One effective way to test for this issue is with vocal tracks. Studio recordings typically place vocalists dead center. You can determine whether the singer's position shifts by playing a low male vocal, a midrange female vocal, and an upper-midrange female vocal. If any of these images are not centered, adjustments to the right speaker are required to properly align the voices. Figure 11 illustrates the correct, centered position of the three singers, along with an example seen in Figure 12 of what you might hear if the right speaker is out of position.

The goal is simple: ensure that every voice and instrument maintains its correct location—regardless of frequency. By mastering the ability to center a phantom image, you take the first major step toward creating a soundstage that is not only accurate but also emotionally compelling.

Figure 11: Solid, Centered Images Independent of Frequency

Figure 12: Images Pulled Left as Frequency Increases

Part 2 — What You Need Before You Begin

- Digital level (e.g., DigiPas)
- Painter's tape
- Sharpie
- Measuring tape
- Your listening chair
- Three reference tracks:
    - » Yello — "Till Tomorrow"
    - » Beethoven — First Piano Concerto
    - » Shostakovich — First Symphony

Part 3 — Pre-Calibration Setup

This set of exercises will help you quickly determine which direction the right speaker needs to move to center a phantom image. Before touching the right speaker, you'll need to complete these three foundational steps:

1. Level the speakers
2. Mark the centerline
3. Establish consistent reference points

Each step builds the accuracy required for the later listening exercises.

Step 1: Level the Speakers

Start by leveling each speaker.

1. Place a digital level (such as the DigiPas) on the **left** speaker. Note the **rake** (tilt front-to-back) and **azimuth** (tilt side-to-side). It's not unusual for a speaker to be significantly off—often by 0.2 to 0.5 degrees in either direction.
2. Return to the listening chair and play the first 30 seconds of *Shostakovich's First Symphony* a few times until you are familiar with the sound.

3. Go back to the left speaker and level it carefully, both front-to-back and side-to-side.
4. Return to your listening position and play the same passage again. Notice any changes? If the speaker was more than 0.1 degrees off, the improvement in sound should be dramatic.
5. Repeat the same process for the **right** speaker.

When there's a significant level mismatch between the two speakers, the soundstage becomes disorganized—some images may pull left, others right. Proper leveling clarifies the sound and its organization, making it much easier to determine the correct direction to adjust the right speaker.

Step 2: Find and Mark the Centerline

Next, identify the room's centerline—or more specifically, the intended centerline between the two speakers.

1. Use the center of your listening chair as a reference point.
2. If your room is symmetrical left-to-right, align the chair's center with the actual centerline of the room.
3. Try to be as precise as possible. While it's true our heads move slightly as we listen, **precision still matters**. Think of the centerline as the **average position** your head occupies during critical listening.

Place a strip of painter's tape on the floor between the speakers, with **two or three marks about six inches apart** indicating the room's left-to-right center point. All adjustments to the right speaker will be made with respect to this line of points.

Step 3: Mark Speaker Position with Painter's Tape

Now mark the speaker positions using painter's tape to ensure **consistency and repeatability.**

1. Measure the distance from the centerline to the **front inside foot of the left speaker**. This distance is labeled **"d"** in the presented diagram.
2. From the centerline, measure the exact distance **d** to the right and place a 6-inch strip of painter's tape on the floor.
3. From the **front edge** of this tape, measure **3 inches forward** and mark the spot with a Sharpie.
4. Place another strip of painter's tape **perpendicular** to the first, starting at this Sharpie mark.
5. From that point, measure **3 inches outward** and place a second Sharpie mark.

This final mark is where you'll place the **front inside foot of the right speaker**. This position intentionally places the speaker *3 inches too far forward* and *3 inches too far to the side*, giving you a clearly "wrong" starting point for the adjustment exercises that follow.

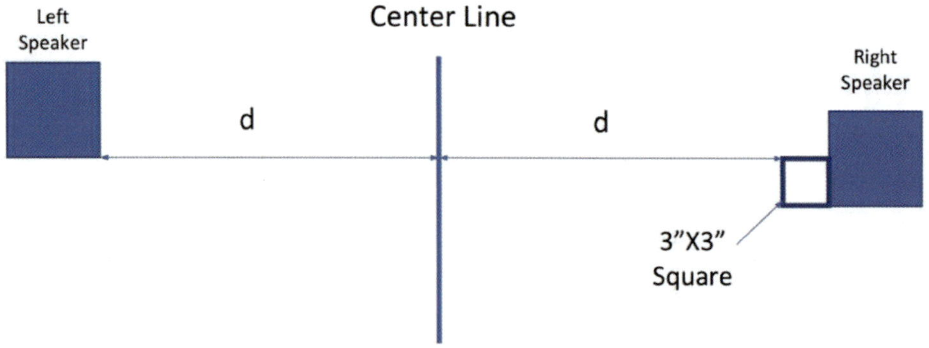

This exercise is *not* about precision. For now, I want you to make large, obvious movements so you can clearly hear how speaker placement affects each song's spatial cues and timing information. We will move to fine-grained precision later.

Begin with the right speaker's front foot both forward and outward. As mentioned, place the speaker at least 3 inches farther out than the center line and 3 inches too far forward. Match the left speaker's current toe-in.

Part 4 — The Reference Tracks and What They Reveal

Different recordings expose different spatial and temporal errors. No single track reveals everything, so for this exercise we will use three songs, each illuminating different details:

- **Yello — "Till Tomorrow"**: Centered pulse; low- and high-frequency symmetry
- **Beethoven — First Piano Concerto**: Large orchestra for left/right timing cues
- **Shostakovich — First Symphony**: Large orchestra for center fill information

"Till Tomorrow" delivers a wealth of information: a strong, centered low-frequency pulse; symmetrical high-frequency clicks; and a centered trumpet. The Beethoven and Shostakovich selections provide complex orchestral soundstages with large masses of instruments spread across the stereo field, which we can conceptually divide into left and right halves. We will break these recordings down and use each sonic cue to determine the proper movement needed to center the phantom image created by the right speaker.

Part 5 — Learning to Feel Sound Pressure

Before you begin listening critically, we need to discuss sound pressure. Speaker drivers push air and create pressure waves that your body can physically feel. You need to become skilled at sensing and interpreting these pressure cues. "Till Tomorrow" is the ideal track to develop this ability because its pulsing bass produces a strong, unmistakable pressure sensation.

Play the opening of Yello's "Till Tomorrow." You will hear the bass line, but more importantly, you will feel the bass pulses. Focus specifically on where on your body you feel this pressure. Replay the intro as many times as needed until you can clearly identify which areas of your body and head experience pressure—and which areas do not.

Shade the areas on the diagram below where you feel pressure. I've included an example on the right from a recent system. You'll notice the pressure is much stronger on the left side, which is incorrect. The pressure should feel even and centered across your torso.

Each time you play "Till Tomorrow," note precisely where on your torso you feel each pulse of pressure.

Figure 13: Torso Pressure Map

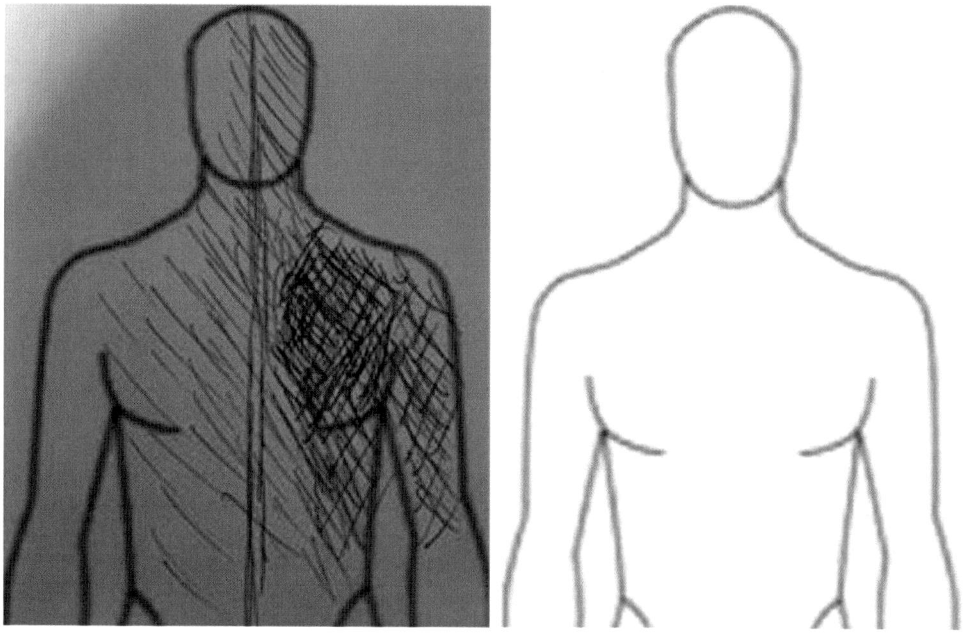

What You Should Hear

In "Till Tomorrow," the pressure you feel should be roughly the size of a dinner plate—centered, solid, and directed straight into your torso. The phantom image of the bass pulse should be perfectly centered. You will also hear three pairs of high-frequency clicks, each symmetrically placed around center, and one centered click. Together, these elements form a triangular pattern. Around the one-minute mark, the trumpet's phantom image should also be centered, with no diffuse trail extending to the left or right.

In the opening seconds of *Beethoven's First Piano Concerto,* the orchestra plays the thematic rhythm: "Baaa... Bup, Bup, Bup."

For this exercise, divide the instruments mentally into two masses—left and right. Do not focus on individual instruments; listen to the collective weight and timing of each half. Several different sections repeat this rhythmic pattern during the first 60 seconds. Your task is to determine whether both halves strike together or if one side lags. If you hear something like: "Ba-baaa... Ba-bup, Ba-bup, Ba-bup," the timing between left and right is off. The correct presentation is both halves playing those notes simultaneously.

In *Shostakovich's First Symphony*, focus on the woodwinds between 28 and 35 seconds. This passage should sound connected, continuous, and centered between the speakers. If the right speaker is positioned too far outward, the orchestra will split into two disconnected halves. If it is too far inward, the center will feel congested or overly dense. Because separation is usually easier to hear than congestion, use this as a guide:

- If you hear the orchestra pulling apart, the speaker is too far out.
- If the center feels crowded, nudge the speaker outward until a hint of separation appears, then move it **back inward slightly** to restore cohesion.

This refined center fill is your indicator that the right speaker is approaching its correct lateral position.

Part 6 — Listening Exercises

These exercises use three independent signals to triangulate the correct right-speaker position:

- Front-to-back adjustment → revealed by Beethoven
- Side-to-side adjustment → revealed by Shostakovich
- Toe-in adjustment → revealed by Yello trumpet + upper mids

Complete them in the order presented.

Exercise 1 — Initial Observation of Right speaker: 3" Forward and 3" Outward

Yello — First 65 seconds. You will observe:

- The bass pulse, which should occupy a tight, dinner-plate-sized circle, is instead diffuse and smeared between the speakers.
- The pressure is strong only on the right side.
- The clicks and pops are audible only on the right; they are missing on the left.
- The trumpet's image appears to move diagonally from front-right to left-rear, rather than progressing along a horizontal line from right to left.
- If your toe-in is matched well, the trumpet at the 1:00-minute mark should still appear centered.

Next, play Beethoven — First 60 seconds. You will hear:

- The right side is ahead in timing, especially on the staccato notes.
- Instead of a unified "Baaa… Bup, Bup, Bup," you will hear something closer to "ba-da, ba-da, ba-da" as the right side fires early.

Finally, play Shostakovich — Seconds 28–35. You will hear:

- The ascending scale is split into two separate "pillars" of sound—one just inside the left speaker, the other just inside the right speaker.
- Although some sound exists in the center, it lacks density and cohesion, having been pulled apart by the incorrect lateral position of the right speaker.

Exercise 2 — Right Speaker -Adjusting Front-to-Back Alignment

Starting from the marked position (3 inches forward and 3 inches outward from center), begin pushing the right speaker straight back in 1-inch increments. After each adjustment, replay the tracks and observe how the pressure, phantom image position, and timing change.

As the speaker moves backward:

- The pressure from the initial bass pulse in Yello's "Till Tomorrow" will begin to shift toward the left side.
- The initial bass pulse will become smaller, tighter, and more punchy.
- The high-frequency clicks will become audible on the left side, restoring symmetry.
- In *Beethoven's First Piano Concerto*, the two halves of the orchestra will time-align as you approach the correct front/back position.
- In *Shostakovich's First Symphony*, the woodwinds will remain spread apart, indicating that lateral (in/out) alignment still needs refinement later in the process.

Exercise 3 —Right Speaker — Adjusting Side-to-Side Position

Return the right speaker to the starting position: 3 inches forward and 3 inches outward. Refresh your auditory memory—you should hear exactly what you heard during the initial impressions. Now begin adjusting the speaker inward, moving it closer to the centerline in 1-inch increments. Continue until the right speaker's front inside foot is at the same distance ***d*** from the centerline as the left speaker.

As you move the speaker inward:

- The pressure will remain strong on the right side of your torso.
- The phantom image of the bass pulse stays pulled to the right.
- The left half of Beethoven's orchestra continues to sound late.
- Shostakovich's center image begins to close up, becoming tighter as the speaker approaches the centerline.

Exercise 4 —Right Speaker — Adjusting Toe-In

Place the right speaker in its home position (not 3 inches forward or outward), but with significantly more toe-out than the left speaker. You should see noticeably more of the right speaker's inside wall than the left. Listen to the trumpet at the **1:00 minute** mark of Yello's "Till Tomorrow." Gradually toe-in the right speaker in small steps until it matches the toe-in of the left speaker. Then continue past that point so the right speaker has more toe-in than the left.

Too much toe-out:

- The trumpet pulls to the left, especially the breathy articulation at the top.
- The soundstage rotates left, making the left side of the orchestra feel more forward.
- The bass pulse is centered in pressure, but its directional vector runs left-to-right.

Too much toe-in:

- The trumpet pulls to the right, again most noticeable in the breathy upper registers.
- The soundstage rotates right, making the right side of the orchestra feel more forward.
- The bass pulse remains centered, but the directionality goes right-to-left.

Part 7 — Practical Advice

I hope you have taken the time to work through the process outlined above. You can read the descriptions and intellectually understand how each movement affects the sound, but there is absolutely no substitute for hearing these changes with your own ears. You need a confident, almost instinctive sense of how the right speaker sounds when it is too far forward or back, too far outward or inward, or has too much or too little toe-in.

- Do the entire process repeatedly. Repetition sharpens your ability to hear subtle differences.
- If possible, have someone gently push the speaker while you stay seated—this reveals changes instantly.
- Experiment with intentionally "incorrect" positions (too far back, too far in, too much toe-in).
- These contrasts teach your ears faster than anything else.

Summary: Establishing the Right Speaker Reference

Through this process, we have used three independent signals to determine the correct symmetric location of the right speaker. Achieving this symmetry is essential because it establishes the reference point that allows the stereo system to form a stable phantom image at the center of the soundstage.

By carefully working through these exercises, you have learned how small adjustments in speaker position influence timing, tonal balance, and spatial stability. More importantly, you have begun training your ears to recognize the subtle cues that reveal when a speaker is correctly aligned.

If you have completed the exercises and properly positioned the right speaker, the improvements should already be clear. The presentation should feel more dynamic, coherent, and transparent. Instruments and voices should appear more stable in space, and the system should begin to sound less like two speakers and more like a unified acoustic event.

We will refine these qualities further when we introduce the left speaker and complete the stereo geometry. For now, take a moment to listen and appreciate the transformation created by a single properly aligned loudspeaker.

At this stage, your right speaker should now be:

- Correctly aligned in time
- Correctly aligned in distance
- Correctly aligned in toe-in
- Producing stable, centered images across the frequency spectrum
- Delivering significantly improved coherence, separation, and transparency

With this reference now established, we are ready to bring the left speaker into the system and complete the stereo relationship.

The next chapter begins that process.

CHAPTER 8

LEFT SPEAKER POSITIONING

"The only way to get a thing done is to start to do it, then keep on doing it, and finally, you'll finish it."

—Langston Hughes

This chapter establishes the left speaker as the reference loudspeaker—the anchor point that everything else in the system must eventually match. The work here can feel slow, but it is foundational. If you are careful and disciplined now, the right speaker will be far easier to integrate later. If you rush the left speaker, the entire system will cap out early—no matter how much you "fix" the right side.

Room Stability and a Practical Warning

Before we begin, remember the single most important condition: throughout this chapter the right speaker remains pushed back and rotated approximately 45 degrees. It must continue to play, but it must not be pointed directly at you. This is not a gimmick. It is the reason the left speaker can be dialed in with clarity and confidence.

It is perfectly acceptable to stop for the day and return later. In fact, fresh ears can be an advantage. When you return, work back through the same songs to the point you left off and note what now sounds different.

However, there is one warning that matters more than most people realize: during the setup process, small changes in the room can derail your progress. I once stopped during the final stages of dialing in the left speaker. When I returned a couple of days later, the sound was suddenly horrible. I checked polarity, channels, footers, and connections. Hours passed before I noticed the front corner closet door was slightly open. When I fully closed it, the sound snapped back into focus. During setup, this kind of change can matter. Once both speakers are fully dialed in, the sound becomes far less sensitive to minor room changes—but while you are still building the foundation, keep the room consistent.

A Bird's-Eye View of the Workflow

As with any new process, it helps to know where we are headed. The chapter unfolds in a repeatable progression. First we establish a neutral starting condition (toe-in, basic rake, and stability), then we find the critical speaker–room anchor point (front-to-back "growl" with open midrange), then we confirm lateral coupling to the right half of the room (Shostakovich trumpet), and only after that do we refine height, lateral timing, rake, toe-in precision, and finally micro-timing and azimuth.

If you want the "quick guide" version of what is coming, here it is in plain sequence:

1. As mentioned, the right speaker should be pushed back and turned away.
2. Adjust the spikes so the speakers are as low as possible and perfectly level.
3. Find the correct toe-in angle by listening to Birdy's voice in "All About You".
4. Use "Chant" to find the best front-to-back distance from the front wall. Find the "GROWL".
5. Verify that we have great bass and that the midrange is good for this position.
6. Use Shostakovich's Trumpet to verify that both sides of the room are evenly pressurized.
7. Find the best speaker height using the bass guitar articulation in "Bass and Drum Intro".
8. Play "Like the Weather" to find the best lateral position for timing and balance.
9. Find the best rake angle using "Sleeping Lessons"
10. Use Kat's voice in "Lucky" to find the exact toe-in angle.
11. Use "Lucky" and Beethoven's First Piano Concerto to remove noise and improve timing.
12. Adjust the Azimuth to reduce tweeter/midrange overlap and "straighten" the image
13. Fine adjustments to timing using various songs.
14. Play through the entire list of songs (including songs used for the right channel) and ensure nothing sticks out.

You can think of the chapter as moving from "big placement" to "fine alignment polish." Large moves first. Micro-moves last.

Phase I: Preconditions and Neutral Setup

Ensure the instructions for the initial placement of the speakers outlined in Chapter 6 have been followed.

Before doing anything musical, set the physical baseline. Adjust the spikes so the left speaker is as low as possible while still being perfectly level front-to-back and side-to-side. The speaker must not wobble. A destabilized speaker will destroy the sound and make every later judgment unreliable.

Now confirm again that the right speaker is pushed back and rotated about 45 degrees. It should be stable and level as well, but it must remain out of the way. I understand that this has been stated multiple times. Granted, it is a fundamental feature of this setup system.

Phase II: Finding the Correct Toe-In Window

Toe-in is first because it places the speaker into a neutral, informative condition that reveals what the speaker is capable of doing in your room. The goal here is not "maximum treble" or "biggest stage." The goal is simply to get your ears into the tweeter's clean working window and then choose the most natural, smooth vocal character inside that window.

The picture in the depiction on the following page illustrates a key reality of multi-driver loudspeakers: the sound from the woofer, midrange, and tweeter does not behave the same way as it leaves the cabinet. In most designs, the tweeter has the narrowest dispersion window by far, while the midrange and woofer radiate more broadly. Because of this, toe-in is not simply a matter of "aiming the speaker at the listener"—it is the method we use to place our ears inside the tweeter's usable window and then choose the smoothest region within that window.

If the speaker is toed out too much, your ears fall outside the tweeter's direct-dispersion window. In that case, you are no longer hearing the tweeter's direct output; instead, you are hearing mostly off-axis treble information, which is typically uneven and rough. That roughness is not a subtle effect—off-axis response often contains significant peaks, dips, and phase irregularities, and it can make the top end sound gritty, splashy, or inconsistent.

The first goal, therefore, is to identify the point where your ear *enters* the tweeter window. Once you find that boundary, the second goal is to continue toeing in until you locate the position where the treble is the smoothest and most natural *within* the window. This matters because the tweeter can still sound "rough" even after you first enter its dispersion area. In many speakers, it takes only a small amount of toe-in to smooth out the midrange, because the midrange driver's dispersion is broader and more forgiving. The tweeter is not. The tweeter often remains audibly uneven until you continue toe-in far enough to move past the region of off-axis roughness and into the most linear portion of its radiation pattern.

In practical terms, the toe-in process should not stop the moment you begin hearing more treble. It should continue until the treble stops sounding irregular and becomes smooth, stable, and consistent—because that is the sign that your ears are positioned in the tweeter's best working window, not merely brushing the edge of it.

Using Birdy to Locate the Tweeter Window

Start with Birdy's "All About You." Everything you need is in the first forty-five seconds, so as soon as the time is up, restart the cut. Take a kneeling stance on the floor about one to two feet in front of the left speaker. Begin with your ear a couple of feet toward the inside of the speaker and position your head so your ear height is very close to the tweeter height. You can move your head slightly up and down to find where the music becomes clearest and most balanced.

Press play and begin swaying your body slowly toward the tweeter. As you move left, you will find the moment where you enter the tweeter window. This boundary should be obvious. As you move your head slightly right the treble drops off substantially. Move your head slightly left and you will hear much more treble energy. Now, start from just inside that window and then continue swaying left and right in very small increments, listening specifically to Birdy's vocal tone. You are searching for the spot where her voice is smoothest or least "rough." The window is usually small. Often you can move into and out of it by shifting your head less than an inch.

Once you have found the center of that clean window, look at the speaker's inside side panel (the side facing the room's center) and identify a visual reference. Note how much of that panel you can see from this near-field listening position. Return to the listening chair and look at the speaker from there. Now go back and forth, making small toe-in adjustments until what you see from the chair matches the visual reference you found up close. When you get close, listen again to Birdy from the listening position and refine toe-in by ear. If the toe-in that sounds best from the chair looks slightly different than what you expected, that is fine. In the listening position, the room plays a larger role than in the near field.

A Process Rule That Matters Later

A natural question at this point is **where the speaker should pivot when you rotate it to adjust toe-in**. There are several defensible answers depending on method and context, but for the process outlined in this book, the approach is deliberate and consistent. During this entire method, treat the front inside spike or footer as the planted reference point when rotating toe-in. Keep that point as fixed as possible and rotate around it. The reason will become more obvious as you progress along the setup.

Establishing a Cursory Rake Direction

Before moving forward, you need to do some cursory rake angle adjustments. This is not the point at which the rake needs to be correct or finalized—it simply needs to be explored and biased in the right direction, so the speaker is no longer perfectly level.

While kneeling in front of the speaker during the toe-in exercise, you identified an ear height where Birdy's voice sounded clear and balanced. Recall where your ear was relative to the tweeter at that moment: below it, directly in line with it, or above it. Now visualize a straight line drawn from the tweeter through that ear position and extended back toward the listening chair. Where does that line intersect when you are seated? Adjust the rake so that this projected line moves closer to your seated ear height.

If the line needs to move upward at the listening position, raise the front of the speaker by lengthening the front spikes. If it needs to move downward, lengthen the rear spikes so the speaker tilts forward. At this stage, the goal is not precision. Simply move the speaker enough in the appropriate direction to break out of a perfectly level orientation.

Process Tip: *a brief note on geometry helps clarify what is happening. Rotating a spike clockwise increases its length relative to the cabinet; rotating it counterclockwise shortens it. How much vertical movement a given turn produces depends on the spike's thread pitch. How much that vertical change alters the speaker's angle at the listening position depends on several factors. For instance, a tall speaker with the tweeter and midrange drivers tightly grouped together at the speaker's top positions are at the end of a long lever arm. When pivoting from the front or rear spikes, relatively small changes in spike length have a more significant impact at the ear in the listening chair than the exact spike change would have if the same drivers were in a much shorter speaker.*

There are two ways to change the rake angle —raising the front or lowering the rear (and vice-versa)—and the choice matters. Raising the front pivots the cabinet around the rear spikes, while lowering the rear pivots it around the front spikes. These two pivot points are not equivalent and will produce different sonic consequences, requiring different compensating adjustments later.

How different those consequences are will depend on the speaker's geometry. If the driver array is vertically aligned on a flat baffle and roughly centered, adjusting rake from either end produces nearly identical results. However, in tall speakers where the drivers sit closer to the top, the difference becomes significant. Pivoting around the front spikes creates a short lever arm relative to the drivers, while pivoting around the rear spikes creates a longer lever arm equal to the cabinet depth.

As a result, in deep-cabinet speakers, adjustments made by raising the front typically have a greater perceptual impact at the listening position than equivalent adjustments made by lowering the rear.

Phase III: Finding the Front-to-Back Anchor Point

Now that toe-in is neutral and rake is roughly pointed the right way, the next step is to determine the ideal front-to-back distance from the front wall where the speaker and room work together harmoniously. This step cannot be rushed. The left speaker's front-to-back position is the foundation upon which everything else is built. Take the necessary time with this step—identify as many promising positions as possible within your allowable range. Experimentation is key, as objective and subjective factors will shape the setup's success. Progress may not be linear, so be patient and persistent. Speaker positioning often follows a circular pattern; the speaker may move into a problematic area and later move into an excellent one.

Using "Chant" to Find the Growl and the Candidate Positions

Start with Nils Frahm's "Chant." Go back to the reference song description and refresh what you are listening for. Once familiar with the song, queue it up on whatever device controls the music—laptop, tablet or smartphone. Then bring your control device over to the left speaker and sit on the floor right in front of it. Place the device next to you. Straddle the speaker with one leg on each side and put your hands on the spikes or wherever you can pull the speaker forward in a controlled way. You need to pull the speaker perpendicular to the front wall. (Follow the blue tape.) This process can be done by pushing from behind, but it is more effective from the front because you interact directly with the driver output and can more easily feel when the room engages.

Press play and listen to the first thirty seconds. The first goal, while seated close, is not to analyze notes or treble. The first goal is simpler: <u>make the room move</u>.

As you pull the speaker forward slowly, you are searching for a growl-like physical engagement in the first few seconds. At specific points it will feel powerful and well-defined. This is not a matter of taste. You will objectively feel the air around you and the floor beneath you move. You will feel the cabinet engage. It either does or it does not.

This is important to understand. You are using "Chant" as a musical tool with your audio system to ask one question: can you feel the air in the room moving with that first chord? If the answer is no, keep pulling. When the answer becomes yes, ask the second question: does the air movement feel and sound like a continuous but diminishing note, or does it pump and breathe? If it pumps, it is not a candidate position yet. If it is steady and diminishing, stand up and move to the listening chair.

From the chair, confirm that the growl translates into the listening seat. You should feel it, not just hear low bass. Sometimes the long droning decay feels "stuck" on the left rather than expanding into the room. It can seem to cycle between the speaker and the center. When that happens the left speaker is not coupling properly to the right half of the room, and later you will likely need lateral movement inward. For now, make a note.

If you are working with a smaller two-way speaker, do not assume this tool is invalid. Even though the chord reaches down to roughly 35Hz, smaller speakers still produce enough information to feel when the stand and room engage. The growl may be less intense than with large multi-woofer designs, but the "either it engages or it doesn't" logic still holds.

What Else "Chant" Must Tell You

Once you have a candidate position with growl, you must evaluate several additional musical traits in the first 37 seconds.

The initial transient attack must be quick and clear. After the strike, the droning decay continues through roughly 19 seconds. When the right hand begins quick strikes at about 20 seconds, you should still hear the decay of that first bass note underneath. The midrange notes should be clear and even. You should hear the felt texture of the hammer on each string. At about 29 seconds there is a strike that should be bell-like with a long decay. That decay is followed by a single strike that sustains downward into near silence before the second low chord drops at about 36 seconds.

This is a lot to assess quickly, but it is exactly why "Chant" is so powerful. It tells you whether the position has the potential to work or not. Some traits can be massaged later with tiny movements, but one trait is essentially pass/fail: you must have growl and an open midrange. If the midrange sounds suppressed, like a blanket over the strings, that position is not worth building on. Move on. But before we consider the other features of "Chant," we need to ensure the position has lateral coupling because without that we will need to find another candidate position.

Process Tip: *a practical approach is to mark your painter's tape as you explore. As you pull the speaker forward, you will encounter positions where the first note produces the correct "growl," yet the right-hand piano notes do not sound even or clear. To keep the process organized, I recommend marking each candidate position directly on the* *painter's tape. Every location where you hear and feel the growl gets a plus sign (+). If that same position also delivers a clear, open midrange, add a second plus sign (+ +). If the growl is present but the midrange is suppressed or uneven, mark it as a plus-minus (+ –). This simple notation allows you to quickly identify which positions are genuinely promising and which should be eliminated, making it much easier to compare and revisit the best candidates as you refine the speaker's placement. The picture depicted below illustrates a real-world example.*

Phase IV: Lateral Coupling Check with Shostakovich

Now you must confirm that the left speaker, in this candidate front-to-back position, communicates with the right half of the room. The most efficient test is the opening trumpet in *Shostakovich's Symphony No. 1*.

Switch to Shostakovich and listen to the opening trumpet. It should stand far back and to the right of the right speaker, planted in the right-rear soundstage. When this happens—even with the right speaker turned away—it is a strong sign that the left speaker is sitting in a fundamentally good place. I like to mark the painter's tape with a third + or — or ~, indicating how good the trumpet's position sounds.

There are several ways the trumpet may present itself. In the best case, it is fully planted in the back right and remains there across all of the trumpet's notes. In a workable but less ideal case, it moves around on the right side but stays right. A cautionary case is where some trumpet notes appear on the left soundstage while others remain on the right; this may or may not be salvageable and requires care. A truly bad case is when most of the trumpet is planted left or seems to originate from the left speaker; that indicates the position will not work.

Opening of *Shostakovich's Symphony No. 1*: If the trumpet's position is correct, this is how it should sound. I prefer the trumpet to be even further back, but if you get this from the initial position of the left speaker, it is good.

Figure 14: Opening of *Shostakovich's Symphony No. 1*

Correctly positioned trumpet.

Figure 15: Opening of *Shostakovich's Symphony No. 1*

Variation of this can potentially be made to work.

Figure 16: Opening of *Shostakovich's Symphony No. 1*

This will never work.

The trumpet's location is a reliable indicator of **lateral coupling between the left and right sides of the room**, as well as the **overall stability of the left speaker's position.** When all of the trumpet's notes remain firmly on the right side of the soundstage, the speaker is generally in a workable region, though small lateral bumps and minor rake adjustments may still be required to stabilize its placement. If several trumpet notes appear on the left side of the soundstage, the left speaker must be moved laterally inward. If the trumpet's very first note originates from the left speaker, lateral inward movement is unquestionably required. As a general rule, the number of notes that drift left of center correlates with how much inward movement is needed.

Stability is assessed while making small front-to-back or lateral adjustments. A position is considered stable if the trumpet remains planted in the right rear of the soundstage as these minor movements are made. If, instead, the trumpet begins to wander, bounce, or shift toward the left when the speaker is nudged, the position is unstable. It is sometimes possible to massage the speaker into a transient position where the trumpet appears correct and other elements seem to fall into place, but this indicates the speaker is sitting on a knife's edge. In such cases, even the slightest disturbance causes the sound to collapse. Although you may have found a micro-position where the trumpet is correctly placed, this position offers no tolerance. The problem will surface later when attempting to match the right speaker: you would be forced to find an equally precise knife-edge position on the opposite side, which is effectively impossible in practice.

After moving the speaker laterally inward and achieving a trumpet that is firmly planted in the right rear of the soundstage, the next step is to confirm full lateral coupling. Continue playing *Shostakovich's Symphony No. 1* from approximately 14 to 33 seconds. The sound should expand and evenly fill the space between the speakers. If the presentation instead separates into distinct left and right sources, with a hollow or underdeveloped center, bump the left speaker inward until the woodwinds occupy the entire soundstage evenly and continuously from left to right.

Figure 17: *Shostakovich's Symphony No. 1*

The sound gap right of center is incorrect.

Be sure to recheck the toe-in after any front-to-back or lateral movement, as both adjustments will alter it. Pulling the speaker forward typically requires slightly more toe-in to maintain the correct relationship with the tweeter's dispersion window. Because toe-in directly influences spatial cues, this change will also affect the placement of Shostakovich's trumpet. After making any toe-in correction, return to the trumpet and confirm that its position has stabilized or improved.

At this point, return to "Chant" briefly and confirm that growl, midrange openness, and decay behavior remain intact.

Return to "Chant"

Once you have found a position that produces the growl and has good lateral coupling, it helps to evaluate what "Chant" is telling you in two distinct categories.

1. The first is whether you have both the growl and an open, functioning midrange. That combination is essentially pass/fail. If the growl is present but the midrange remains suppressed—like a blanket over the piano strings—there is very little "massaging" that will make the position fundamentally viable.
2. The second category includes the transient attack, the decay into silence, and the spatial behavior of Shostakovich's trumpet. These elements can often be improved with very small, deliberate movements that slide the speaker into the pocket.

One detail to keep in mind is that the initial strike in "Chant" contains a very quick "bounce" immediately after the transient. This can be seen clearly using Audacity when viewing the waveform of the opening chord as shown in the figure below. When you first land in a growl position, the transient is often slightly blurred, and the bounce may be hard to hear. As you refine the position with a few small bumps, that "bounce" should become audible. Treat it as a sign that the speaker is snapping into a more precise alignment.

Figure 18: Waveform of Initial Chord in "Chant"

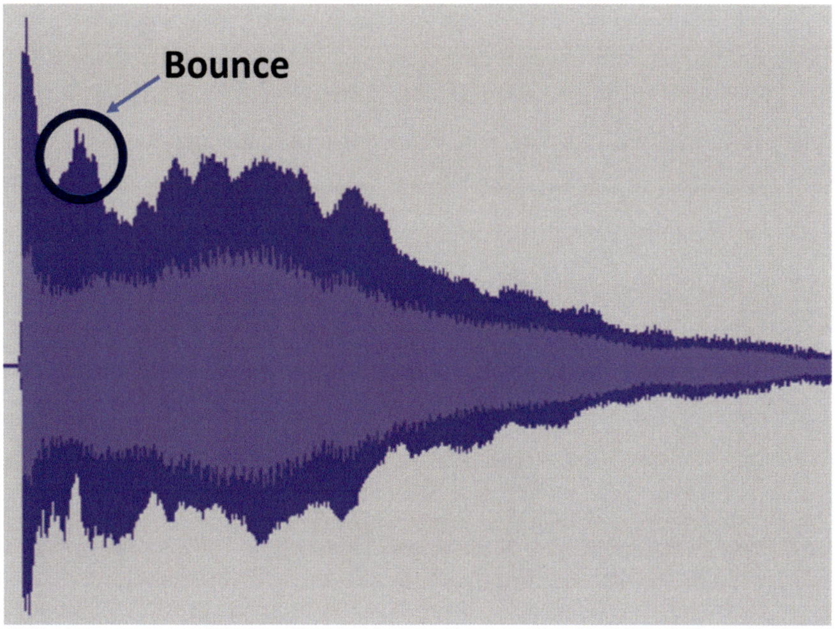

When pulling or pushing the speaker forward, it is common to stop slightly short of the best position—or pull slightly past it—because the growl is easiest to detect once you enter the general region, not necessarily at the exact optimum point. In many near-candidate positions, you may notice the decay after the initial strike oscillating in level. I refer to this as "breathing." The decay should be steady, smooth, and continuously diminishing. If you hear breathing, bump the speaker forward slightly and listen again to see if the decay becomes more stable. If several small bumps forward or backward fail to produce a steady decay, abandon that position and search for another candidate zone.

The same approach applies to the transient and the decay-to-silence. If the transient sounds weak or the note does not decay cleanly into silence, try bumping the speaker forward—not backward—and reassess. Continue only while the transient sharpens and the decay becomes smoother and longer. As the transient and decay improve, you will usually notice Shostakovich's trumpet becoming more securely planted in the correct right-rear position.

If the midrange lacks clarity even though the growl is present, use tiny lateral adjustments. If the trumpet is already planted on the right, experiment with a small lateral bump outward. If you previously had to move the speaker inward to get the trumpet planted, continue working in that same inward direction with very small lateral corrections until clarity and stability improve together.

Although rake has not been a major focus yet, it is worth tracking the piano's vertical placement in "Chant." The piano should appear at or slightly above eye level. If it is too low, rake the speaker back slightly to raise the image. This also nudges you closer to the eventual correct rake region. After any rake change, return to the trumpet and confirm it remains planted; a correct rake adjustment often helps stabilize the trumpet's position rather than disrupt it.

The importance of finding the best starting position cannot be overstated. This is the foundation on which everything else is built. If the foundation is unstable, the final result will be unstable. Take your time. Learn what each change does. Experiment with pulling the speaker forward and listening to how the growl, transient attack, decay behavior, silence, and midrange clarity change from position to position. It is entirely reasonable to spend an entire Saturday afternoon—or more—learning how "Chant" behaves in your room and how to manipulate it. That time pays dividends throughout the rest of the process.

By this point, you should have explored most of the usable front-to-back range and marked multiple candidate positions on your painter's tape. Move quickly among your best two or three positions and compare them. Once you choose the strongest one, massage the speaker 's position until all of "Chant's" key markers are satisfied: growl, clean transient attack, clear midrange, smooth decay, and real silence. When those elements line up, you have reached the position where the speaker and the room cooperate.

If you stopped here and simply positioned the right speaker symmetrically, the sound would already be dramatically improved compared to a random or formula-based placement. But we are not finished. It is almost certain that the speaker's height still needs adjustment to improve bass clarity and articulation.

Before moving on, take a moment to play The Shins' "Sleeping Lessons" and form an initial impression. We will use it later, but it is important to hear how unruly it sounds right now. At this stage it should resemble a series of "bubbles" emanating from the left speaker, surrounded by excessive bloom, resonance, and ringiness—often so much that it can feel like it is inflating the ceiling. That reaction is normal here. As the left speaker moves closer and closer to correct, the bloom will steadily diminish, and the bubbles will become clearer and more transparent. Remember what it sounds like now. As you progress through each subsequent refinement step, check back periodically and notice how the track cleans up. Later, once the system is much closer to alignment, we will use this track for timing work. For now, the speaker is still too far from that level of refinement to make it a primary tool.

Phase V: Speaker Height—Bass Articulation and Weight

Once you have a stable candidate front-to-back position with good coupling cues, height becomes one of the easiest and most impactful changes you can make. Height is not some cosmetic tweak. It can transform bass from foggy and stuck-on-the-floor to articulate and rhythmically meaningful.

Use *Poem of Chinese Drum* first to establish a reference for bass fullness. The large drum should have a deep BOOM, not a pitched-up boom that sounds like bongo drums.

Then switch to Nils Lofgren's *Bass and Drum Intro.* Focus on three questions:

1. where does the bass seem to originate in space?
2. how distinct is each note as he plays?
3. how clean and intelligible are the low notes around 1:23?

Raise the speaker by turning all four spikes upward by a quarter turn. Listen again. Repeat in quarter-turn increments until the bass becomes clear and articulate while still retaining weight and impact. Use *Poem of Chinese Drum* to ensure you are not leaning the bass out too much and use *Bass and Drum Intro* to ensure articulation is improving.

You will eventually reach a point where the bass becomes too thin. At that point you can either come back down an eighth of a turn and zero in, or, if you have enough spike length, continue upward until you find a second zone where bass is both full and articulate. Many speakers do not have enough thread length to reach the second zone, so do not assume it will be available.

Be aware that being slightly too high or low can induce high-frequency noise. The upper register notes may become fuzzy or unclear. If that happens, lowering slightly often reduces the noise. In the grid model, you are snapping onto the correct height grid line.

In *Bass and Drum Intro*, pay attention to the section from roughly 0:23 to 0:52 where Nils intentionally pulls his thumb off the strings. As height rises, the attack can become exaggerated or aggressive. If it becomes too aggressive, a small bump forward often helps.

Also note the "3-band chord" around 0:52–0:58. It should sustain evenly without one band—treble, midrange, or bass—bending independently. If the bass band bends low, height is likely still wrong. If midrange and bass bend together, you may need a combination of height and rake back later.

Following the "3-band chord," the sound should drop to dead silence before he starts playing again. Once he starts, we should be able to hear the crowd cheering behind him. This should be very "alive" sounding.

When you believe height is correct, return briefly to "Chant" and confirm growl and midrange did not collapse. Height changes can subtly shift other relationships.

Phase VI: Lateral Timing and Driver Coherence with "Like the Weather"

Now we must find the lateral position that makes bass, midrange, and treble play together in time. This is where time coherence and "boogie" (otherwise known as PRAT) begin to emerge. For this step, we use the music tool "Like the Weather" by 10,000 Maniacs. Although this song is not an audiophile showpiece, it contains exactly the characteristics we need for setup. It features five instruments—a bass guitar, two guitars, a snare drum, and a keyboard—playing together across a frequency range that spans bass, midrange, and treble. In addition, the recording has a very low dynamic range, scoring only 4 in the Roon utility. This score means that nearly all of the instruments are playing at roughly the same loudness. While this is not usually a desirable musical trait, it is extremely useful here because it forces multiple instruments across the spectrum to compete equally for timing coherence. Finally, the song opens with a syncopated introduction. Syncopation is fragile and unforgiving; even small timing errors cause the music to fall apart. If the timing is right, the band sounds like the professional group it is. If it is not, the groove collapses immediately.

As the song plays, there are several things you should notice. The snare drum should have a sharp, well-defined snap. The keyboard should be clearly audible rather than buried or smeared. The bass guitar should exhibit both rhythm and texture. All of these elements matter, but it is impossible to establish the syncopated rhythm without a properly articulated bass line. By this point, the speaker's height should already be close to correct, and the bass should be articulate. If it is not, you may need to make a very small height adjustment—typically on the order of one-sixteenth to one-eighth of a spike turn. Raising the speaker slightly will usually improve bass clarity, while lowering it can add a bit more bass weight.

Only the first 19 seconds of the track are required, which contain the syncopated introduction repeated twice. You will almost certainly need to play this section multiple times to internalize the feel. It is strongly recommended that you listen to this song on headphones before beginning, as headphones present the syncopated rhythm without any room interaction or driver-integration issues. When assessing the left speaker with this track, I allow my head and torso to move subtly with the beat to feel the rhythm physically. As the speaker's timing improves, the syncopation becomes easier to follow. When the speaker is far from its correct position, it can be impossible to get into the song at all. After some lateral movement, the rhythm begins to emerge. When the position finally locks in, the change is unmistakable: the song comes alive and develops what I describe as "boogie."

Some listeners naturally connect with this feel more easily than others. If you find that you do not, rely on your ears instead of your body. Focus specifically on the relationship between the snare drum and the bass guitar. Many of the snare hits should land directly on the bass beat. As you move the speaker laterally, you will hear the snare arrive either slightly early or slightly late relative to the bass. When those two elements align, you are very close to the correct lateral position.

Once you can perceive the syncopated rhythm clearly, assess whether the band sounds like it is playing together in time. In most cases, the speaker will not yet be in the correct position. The amount of movement required depends on how far out of time the band sounds, but it typically falls in the range of 1/8" to 1/4". Since this may be your first time performing this adjustment, begin by moving the speaker laterally outward—away from the centerline—by 1/8" and listen again. If the syncopation improves, continue in that direction. If it does not fully lock in, move the speaker outward again by about 1/16" and reassess. You may need to repeat this process several times before the band truly plays together. Once you are close, use very small bumps near the bottom of the speaker cabinet to fine-tune the position until it sounds right. Do not stop until the timing is correct. If you overshoot, bump the speaker back toward the centerline.

You might reasonably ask why the speaker is moved outward rather than inward. In reality, the timing and syncopation could be aligned by moving the speaker in either direction, as multiple alignment zones exist due to the grid behavior discussed earlier. The choice is partly a matter of preference. A wider speaker placement often produces a more open presentation, while a narrower placement tends to sound denser. Another important consideration is sidewall coupling. Lateral movement changes how the speaker interacts with lateral room modes, which directly affects bass character. Pay close attention to how the bass guitar changes as you move the speaker outward or inward and choose the presentation you prefer.

Recall the concept of rotational timing introduced earlier. After you find one lateral position where the timing aligns, you could continue moving the speaker laterally. The timing will initially worsen, but if you continue far enough, it will begin to improve again, eventually revealing another position where the timing once again aligns. This is normal and reflects the periodic nature of grid alignment.

After completing the lateral adjustment, it is essential to return to "Chant" and confirm that the speaker is still in its best front-to-back position. Sliding the speaker sideways makes it very easy for it to drift forward or backward by as little as a thirty-second of an inch. If the low-frequency growl is no longer present, move the speaker back slightly and then pull it forward again until you rediscover the growl and restore the correct front-to-back anchoring.

You must also reaffirm the lateral coupling. Play Shostakovich and listen to the trumpet. If it remains in the rear right of the soundstage then all is good. If the trumpet has moved to the left side, we need to reassess the direction you moved the speaker. If you moved it laterally outward, then you will need to move it inward instead. Use "Like the Weather" to make the necessary adjustments. We must have: the growl of "Chant," the lateral coupling of Shostakovich's trumpet, and the timing coherence of "Like the Weather."

Phase VII: Toe-In Refinement and the "Low Noise Window"

Now that front-to-back, coupling, height, and lateral timing are close, toe-in must be reset and refined with greater precision. Any lateral movement will have altered the toe-in, so it needs to be reset to its baseline orientation. If the speaker was moved outward, increase toe-in slightly. If it was moved inward, reduce toe-in accordingly. Use the visual reference along the speaker's side panel that you established earlier to get close, then listen to Birdy's "All About You" to bring the toe-in nearer to correct.

For the final adjustment, cue up "Lucky" and focus on Kat Edmonson's voice. Move only your head in small increments to the left and right of center. You will find a position where her voice sounds clear and vibrant. Adjust the speaker's toe-in or toe-out so that this optimal point aligns with your normal, relaxed listening posture. Next, move your head in even smaller increments. You will likely discover a very narrow position where the noise in her voice drops noticeably, and the presentation becomes quieter and more settled. That is the target.

Make the final toe-in corrections using tiny physical adjustments. Hold the front, inside spike or footer firmly with one hand to keep it stationary. With the other hand, lightly bump the back inside corner of the speaker to increase toe-in, or bump the outside rear corner to reduce it. These movements should be extremely small; they may feel insufficient, but they are not. A minute rotation at the speaker translates into a much larger positional change at the listening seat, so restraint and precision are essential here.

Phase VIII: Transitioning to Smaller Adjustments and Removing Sliders

With the major movements complete, it is time to remove the speaker from the sliders and place it directly on the floor. If you are not using a Lo-Pro jack or air bladder, it is best to have someone assist you. Tilt the speaker carefully and deliberately, making sure the sliders do not shift underneath it. Placing small pieces of tape on the floor to mark the exact position of the speaker is a good precaution in case anything moves slightly during this step.

Before doing this, however, you must record two critical reference values: the height of the speaker at the front and its rake angle. The sliders add height to the speaker, so once the speaker is placed directly on the floor, these values must be recreated precisely. How you measure depends on your floor. On carpet, use a flat, stable reference surface such as a bamboo cutting board or a piece of MDF (Medium-Density Fiberboard) to establish a consistent measuring plane. Measure the height at the front center of the speaker with a ruler or use an adjustable gauge or reference block placed beneath the cabinet. After the speaker is off the sliders, adjust the spikes until the front height and rake angle match what you measured while the speaker was still on the sliders.

Pro Tip*: If the speaker is sitting on a hard floor, an especially precise method is to use a flat-head Allen bolt and coupler, both readily available at a hardware store. Wrap a small amount of Teflon tape around the threads, connect the Allen bolt to the coupler, and place the coupler on the floor beneath the speaker. Unscrew the bolt until its head just touches the bottom of the cabinet, then secure the bolt and coupler in place with tape. This creates a highly repeatable height reference that can be matched exactly once the speaker is resting on the floor.*

After the speaker is back in position, listen again to all of the music tools used so far—*Chant*, *All About You*, *Lucky*, *Like the Weather*, *Bass and Drum Intro*, and *Poem of Chinese Drum*. Confirm that the growl is still present and that bass articulation remains intact. If either has changed, make very small adjustments to the front-to-back position and/or height to restore the speaker to its original alignment before it was removed from the sliders. These corrections should be minimal. When everything is right, the bass will sound noticeably more solid. This is often easiest to hear in the initial piano strike of *Chant* or in the drum hits of *Poem of Chinese Drum*. The added solidity comes from the speaker being firmly coupled to the floor rather than floating on the sliders.

Phase IX: Refining Rake with "Sleeping Lessons"

While adjusting the rake angle by rotating the rear spikes, it is very easy to turn one spike slightly more than the other. After repeated adjustments, this imbalance can become significant enough to introduce a sideways tilt, unintentionally changing the **azimuth angle**. Because azimuth directly affects how the drivers integrate, it must remain constant while rake is being adjusted; otherwise, your perception of the rake change will be skewed by an unrecognized azimuth error.

If your speaker has a flat top surface, place the Digipas level on top of the cabinet while performing these adjustments. This allows you to check azimuth quickly and repeatedly, ensuring it remains at 0.000 degrees as the rake is altered. If the speaker does not have a flat top, you must establish a reliable reference plane for the level. Regardless of the method, check azimuth frequently and consistently to ensure it does not drift as you work.

Pro Tip: *An accurate level is a valuable tool. It provides confidence that something is truly level or plumb. But a high-precision level will reveal the difficulty in positioning objects exactly where you want them. The more sensitive the instrument, the more precise you must be in your measuring process.*

I've discovered that one of the most overlooked pitfalls when using a level is the impact of your own body weight on the floor. If you want to know the exact position of the object you're leveling, you need to step away and observe the reading without the influence of your weight. This isn't possible with a bubble level—unless you hang a camera above it, and then parallax becomes a problem. The only practical solution is a level with Bluetooth connectivity that displays the reading remotely.

The best way to understand this is to try it yourself. Using the Digi-Pas DWL1500XY machinist level, we've included four screenshots of the Digi-Pas app on my phone. The level was placed on top of a flat speaker, aligned with the cabinet's front baffle.

1. *Seated in the listening chair: A neutral azimuth of 0.000° and a rake angle of 0.060° back.*
2. *Standing 12" in front of the speaker: The azimuth shifts to 0.005° tilted counterclockwise, and the rake angle decreases to 0.050° back.*
3. *Standing 12" from the left side of the speaker: The azimuth shifts to 0.015° tilted counterclockwise, and the rake angle reads 0.057° back.*
4. *Standing 12" from the right side of the speaker: The azimuth shifts to 0.009° tilted clockwise, and the rake angle reads 0.067° back.*

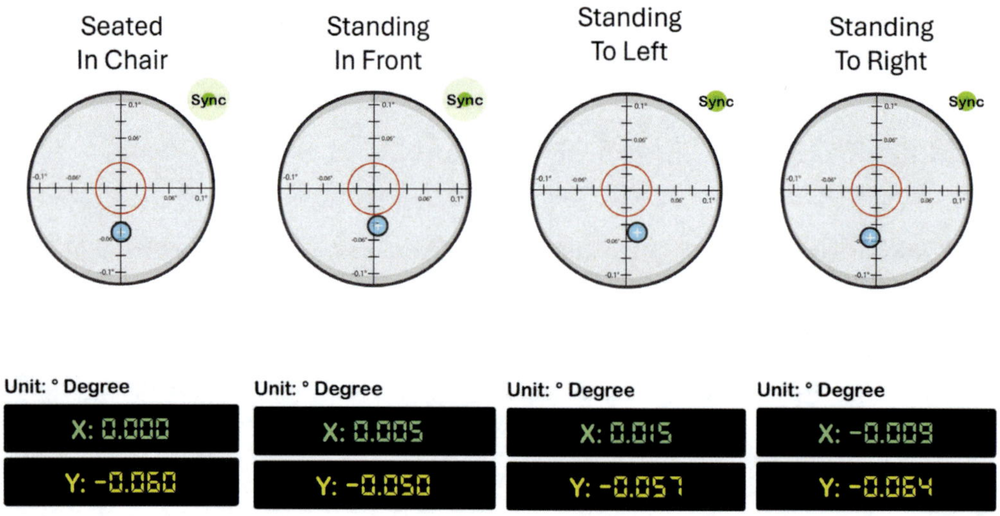

Standing near the speaker clearly alters its position in space, well outside our desired tolerance for a precision setup. This particular floor is actually quite stable compared to many I've encountered over the years. If this were a heavy turntable on a stand, the impact would be more severe because of the elevated mass.

While expensive, a high-resolution Bluetooth-enabled level is the only tool I know of that allows you to accurately determine the true position of a speaker in space—or the level of a turntable—without the influence of your body weight. However, placement of the device is also a critical consideration. Surprisingly little placement deviation will produce inconsistent readings. A level like the Digi-Pas DWL1500XY reads to 0.000° with a calibrated resolution of ±0.001°. To put this in perspective: a tilt of 0.100° equals a 1.74 mm rise over one meter; 0.050° equals a 0.87 mm rise over one meter. It doesn't take much angular deviation to move the drivers out of coherence, given the complex frequency relationships we're working with in music.

Speakers are consumer products, not machinist's tables where absolute flatness is guaranteed. Find the flattest spot on your speaker and place the level in exactly the same position every time. Mark it with blue tape so it's easy to locate again—just don't place the level on top of the tape. We are not trying to place the speaker in an "absolute" position with a specific readout. We want a consistent reference we can use to carry out precise positional changes. There is only so much we can do to ensure precision, and what we measure will never be truly accurate to 0.001°—but we can get as close as possible to consistent and trust our brain to tell us the rest.

Many speakers lack a flat top, making an essential reference plane for accurate measurement. In most designs, the bottom of the cabinet is flat. A bamboo board or a piece of MDF can be held firmly against the bottom of the speaker to create a temporary shelf for the level. When using this method, consistency is critical. You cannot overcome the effect of your weight—sit in the same position, hold the board in the same position each time and apply the same pressure so that readings remain repeatable.

Continuing On

Earlier in the process, you positioned yourself in front of the speaker and used Birdy's *"All About You"* to identify a location where the sound became quiet and natural, then began nudging the rake angle in the appropriate direction. You likely continued refining this using the piano's vertical placement in *"Chant."* Now it is time to refine the rake angle further using *"Sleeping Lessons"* by The Shins.

The following illustrations depict several "bubbles" in front of the speaker, surrounded by jagged lines. The bubbles represent the electronic pulses in the song as they rise and fall in frequency. The jagged lines represent bloomy noise generated by misalignment. When you listened to this track earlier in the process, it likely resembled the left illustration: the bubbles were smeared together, accompanied by a great deal of bloom and ringiness. At this stage, because much of the room's influence has already been windowed out, the track should now sound significantly cleaner.

As you listen, visualize each tone or pulse as an individual bubble of sound. There should be clear space between one bubble and the next. Sometimes the entire stack of bubbles—from low to high frequency—can sound fused together; in other cases, only the top few bubbles remain joined. The lower bubbles correspond to lower frequencies and separating them usually requires a small **height adjustment**. The upper bubbles are higher in frequency and are separated primarily through a very small **rake adjustment**, typically raking the speaker back.

If you consider the geometry, raising the speaker's height and then raking it back by lowering the rear spikes is functionally equivalent to simply raising the front spikes. Use this relationship deliberately. If the lower bubbles are blurred together, raise the front of the speaker until they separate. Once the lower bubbles are distinct and only the top two or three remain joined, lower the rear spikes slightly to rake the speaker back. This will separate the remaining bubbles and make them individually distinct.

When you achieve clear separation between the bubbles, the rake angle is very close to correct. At this point, the top end will sound more open, the bloom will be reduced, and the treble will become smoother and more pleasant.

Image Height as a Preference, Not a Panic

The rake adjustment directly determines image height. Listen to a female vocal track and focus on where the singer appears vertically in space. Does her voice seem to originate at the midrange driver, at the tweeter, or above the tweeter? As the rake angle of the left speaker increases, the perceived image height rises accordingly. This is ultimately a matter of personal preference: some listeners favor a lower image, while others prefer a higher one. You are in control of this placement.

This is also an area where system noise can be misleading. Excessive noise often causes the image to appear artificially elevated. For example, during a recent setup, the image height initially appeared to be at the top edge of the tweeter. After installing several devices in the streamer and DAC—substantially lowering the system's noise floor—the image height dropped to the bottom of the midrange driver. The change was not caused by speaker movement, but by noise reduction revealing the system's true spatial behavior.

Azimuth

At this point, you may notice that five of the six available degrees of freedom have been addressed: front-to-back, lateral position, height, toe-in, and rake. The remaining degree of freedom is **azimuth**, or sideways tilt. It is important not to adjust azimuth yet.

If azimuth is changed too early, later refinements can make that adjustment unnecessary—or worse, cause it to become an overcorrection. Consider a hypothetical example. While listening to a song, you notice what seems to be an azimuth issue and make a small outward adjustment of 0.003 degrees. The problem appears resolved. You then continue refining the setup, moving the speaker slightly forward and outward, followed by a small rake adjustment and a compensating toe-in change. Listening again, you believe another azimuth adjustment is needed and apply an additional 0.007-degree change. Everything now seems aligned, but in reality, you have likely overcorrected. Had you returned the azimuth to 0.000 degrees, you may have found that only a 0.001-degree adjustment was actually required.

What has happened is that, although the grid lines are aligned, the speaker has effectively shifted one grid too far laterally. This example illustrates why azimuth should be adjusted only after all other movements—bumping, nudging, rake refinement, and toe-in correction—are complete.

A Decision Point

You have now made the major adjustments necessary to establish the left speaker's home position. At this stage, the sound should already be quite good. You have reached a decision point: either continue refining the left speaker or move on to setting up the right speaker.

To help make this choice, listen to several familiar recordings. If you hear specific issues that still need to be addressed, continue working with the left speaker. If the sound is satisfying overall, you may proceed to the right speaker. If this is your first time performing this process, it is often wise to move on to the right channel sooner rather than later. The further you push the left speaker's performance, the more difficult it becomes to match it precisely on the right.

Phase X: Micro-Timing, Noise Removal, and the Piano Standard

As you listen to a wider range of music, you will almost certainly notice remaining issues—perhaps a lack of transparency, truncated decay, or missed fine timing cues. These are signs that further refinement is needed to improve coherence. From this point forward, adjustments will be extremely small: tiny bumps, light taps, or minute spike rotations.

Beethoven as the Noise and Spacing Reference

To refine cohesion, use three primary music tools: Kat Edmonson's *"Lucky," Rossini's L'Italiana in Algeri,* and *Beethoven's Piano Concerto No. 1.* As sound quality improves, the necessary adjustments become smaller and smaller. You will cycle through these tools repeatedly, tightening the circle with each pass until no further improvement is possible.

True transparency is achieved by minimizing room interactions and precisely aligning the grid lines. This typically requires very small lateral or front-to-back adjustments, along with fine rake refinement. Beethoven's *First Piano Concerto* is particularly valuable here, as it reveals both acoustic noise and timing coherence.

Listen to the first five minutes of the concerto. When the piano enters at approximately 3:19, note its apparent height. It should appear slightly above eye level. If the piano manifests too high, rake the speaker forward slightly. If it seems too low, rake the speaker back. These adjustments should be extremely small.

As the piano plays, the notes should be evenly spaced, distinct, and crystal clear, with a delicate, bell-like quality. They should never sound bloomy or aggressive. It is tempting to blame recordings or performers for smeared articulation, but in this case, that assumption is incorrect. Arturo Benedetti Michelangeli was a world-class pianist with extraordinary control. Any lack of clarity or space between notes is not present in the recording and is certainly not a performance flaw—it is caused by noise or misalignment in the playback system, and it is our responsibility to remove it.

Two illustrations in the following Figures help clarify this concept. The first Figure shows the piano keys at the correct height, evenly spaced and level. The second Figure shows the keys jagged, noisy, and rising from left to right. We have already addressed height; these illustrations serve to visualize how noise and misalignment distort timing and spacing. Our goal is to make the piano resemble the first illustration—stable, even, and unforced.

Figure 19: ***Beethoven's Piano Concerto No. 1*** **@ 3:26**

It should sound like this.

Figure 20: ***Beethoven's Piano Concerto No. 1*** **@ 3:26:**

It should not sound like this.

The second illustration represents the piano keys as a jumbled mess. Some notes overlap, some jump out too loud while others seem too soft, and a layer of noise—especially in the higher frequencies—makes the whole presentation feel jagged and unsettled. Our goal is to "even out" the piano, so it resembles the first picture: level, orderly, and evenly spaced. When the piano sounds noisy and slurred—as though there is not enough space between notes—the primary correction is the speaker's lateral position. If the higher notes sound aggressive or edgy, the correction is typically a small forward bump. Using the illustrations as a mental model, think of lateral movement as reducing overlap between the "keys" and restoring proper spacing, and think of forward movement as flattening the presentation so the keys stop sounding sharp, elevated, or noisy.

Begin by evaluating the piano from 3:19 to 3:34. The passage from 3:31 to 3:34 corresponds directly to the notes represented in the illustrations. At the start of this section, listen closely to the clarity of each note. If the notes sound unclear or slightly muffled, bump the speaker outward and listen again. Continue through 3:34, paying particular attention to 3:31–3:34, because this is where noise and aggressiveness are easiest to hear. If you detect harshness—especially in the treble—bump the speaker forward slightly. After each bump, whether outward or forward, restart the passage and listen again. Repeat as many times as necessary until the notes are clean, evenly spaced, and free of glare. It is completely normal to cycle through this short segment ten times or more while making micro-corrections.

Starting at 3:40, Michelangeli plays two long descending arpeggio runs that span nearly the full range of the instrument. The highest pitch is E6 (1318 Hz) and the lowest is C2 (65 Hz), which means this single musical gesture crosses your speaker's treble, midrange, and bass regions. Ask yourself whether the run sounds like one coherent instrument—a single piano—or whether it sounds like the speaker's drivers are handing off the sound in a way you can perceive. The runs should not sound slurred in any way. Each note must have a clear attack and an audible pocket of space before the next note begins.

The absence of space is an insidious problem because it can be easy to miss at first. If you are not accustomed to listening for note spacing, you may think, "Yes, I can hear every note." But that is not the question. The question is: how much space exists between notes, and does each note remain distinct from the one that follows? This passage is not meant to sound legato. If the notes blend together, the system is still adding smear and noise, and further refinement is needed.

Beethoven's *First Piano Concerto* (Descending Arpeggio at 3:40)

If you still hear a lack of spacing between notes or a general indistinctness, go to the speaker and give it a very small bump at the center of the bottom inside edge so that the cabinet moves slightly outward. You might reasonably ask, "Didn't we already do this during the section from 3:19 to 3:34?" We did—but this passage is played faster. The earlier section required enough space for slower notes; this later run demands additional space so the faster notes can remain distinct.

After making the bump, listen again. Pay attention to whether the opening notes of the piano solo have more "twinkle" and whether the notes in the descending run are more clearly separated. Continue this process until the notes sound clean, distinct, and properly spaced, or until further bumps produce no improvement. At that point, stop.

It is possible to reach a plateau where additional adjustments no longer improve clarity, even though the notes are not absolutely perfect. This may reflect a limitation of the speaker design itself or a higher-than-ideal noise floor elsewhere in the system. If you have not already spent time addressing the electronics chain, this is the moment to do so. Noise sources can be subtle and difficult to isolate, and they are addressed in detail in Chapter Three.

Because the speaker has been moved both forward and laterally, compensating adjustments are almost always required. Revisit rake and toe-in. Remember that the piano should appear slightly above eye level; use this cue to refine the rake angle, which is adjusted only with the rear spikes.

For toe-in assessment, listen to the piano from 3:13 to 3:26. You can do this from the listening chair by shifting your head slightly left or right. If, with your head moved a small amount to the left, the piano sounds quieter or less noisy, go to the speaker, hold the front inside spike firmly in place, and give the inside rear corner a very small bump to increase toe-in. Listen again and repeat as needed until you reach the "low-noise window," where the piano sounds cleanest with your head pointed straight ahead. If you prefer, you may also use "Lucky" or "All About You" to assess and fine-tune toe-in, just as you did earlier in the process.

Rough Driver Timing Using Drum and Xylophone Cues

Once the acoustic noise is reduced enough to hear timing cues, you can begin aligning drivers. We must continue refining the speaker's performance to achieve a truly cohesive and coherent sound. At this stage, the speaker has been positioned to minimize noise, which allows us to begin evaluating **inter-driver timing**. It is very likely that the bass, midrange, and treble are not yet arriving at your ears at exactly the same moment. I use the term "integrated" deliberately, because what you hear is a combination of direct sound from the speaker and reflected sound from the room's boundaries. The brain naturally fuses some reflections into a single acoustic event while perceiving others as separate. Our objective is to align the speaker so that all of its elements function as one, producing a unified arrival of sound. For that to occur, all frequencies must arrive in sync.

A rough but important first step toward this goal was already taken using "Like the Weather," where we ensured the band's syncopated rhythm locked together. Now we move beyond gross timing alignment and begin refining how the individual drivers relate to one another.

The two illustrations above visualize a single speaker's sound as overlapping "grids." One grid represents the woofer, another the midrange; a tweeter grid also exists but is omitted for clarity. In the left illustration, these grids are misaligned, while in the right illustration they are aligned. With very small physical adjustments—either tiny bumps or minute spike rotations—we progressively bring these grids into alignment from woofer to midrange and from midrange to tweeter. Although the illustration shows the grids as a two-dimensional web, in reality they exist in three dimensions. This means alignment must be considered in each degree of freedom.

Understandably, there are many overlapping grids at play that can be separated conceptually by learning to recognize each one's sonic signature. Correcting timing misalignment requires extremely small movements: very light taps on the cabinet with your knuckles or spike rotations on the order of 1/32 of a turn or less. In some cases, the adjustment may be so small that you are effectively just breaking the thread.

Adjusting the time of arrival between drivers requires deliberate thinking. If needed, revisit the section in Chapter 5 titled "Arrangement of Speaker Drivers." The most straightforward example is a conventional three-way speaker with the tweeter mounted above the midrange. In that case, the relationships are as follows:

- Bass-to-midrange timing is primarily adjusted with height: rising the speaker advances bass timing, while lowering it delays bass relative to the midrange.
- Midrange-to-tweeter timing is adjusted with rake: tilting the speaker forward advances the treble, while raking it back delays the treble.
- Fine treble timing is handled with azimuth: rotating inward advances the tweeter slightly, while rotating outward delays it.

These relationships form the basis for the fine timing adjustments that follow, and they must be approached with patience and restraint, as very small movements can produce meaningful changes in coherence.

Intro Drum Rhythm to "When the Levee Breaks"

We will use Led Zeppelin's "When the Levee Breaks" to assess bass-to-treble timing alignment. The ten-second drum introduction—two measures long—contains unusually clear timing information. The hi-hat plays eight evenly spaced notes while the kick drum and snare land on quarter notes, as shown in the excerpt below. Play this intro several times to learn its internal timing. On the first pass, listen only to the kick drum. On the second pass, listen only to the snare drum. On the third pass, focus entirely on the hi-hat.

Intro Drum Rhythm to "When the Levee Breaks"

Once you can clearly "hold" each element in your ear, play the intro again and evaluate whether the hi-hat is landing precisely on the beat with the kick drum, and whether it is also landing correctly on the beat with the snare. If the hi-hat consistently leads the kick drum, raise the speaker's height by a very small amount. Continue the cycle—listen, assess, adjust height—until the hi-hat and kick lock together in time. If the hi-hat is consistently behind the kick drum, lower the speaker's height slightly until they synchronize.

Next, use Kat Edmonson's "Lucky" to assess midrange-to-treble alignment. In the opening seconds, vibraphone keys are double-struck. Each double strike should land on the same beat, effectively fusing into a single impulse. At first, both strikes are higher in pitch; around eleven seconds in, one hand drops in pitch. Throughout, the two strikes should remain extremely close in time. You should not hear a separated "ba-da, ba-da, ba-da" effect. Determine which component arrives first: the treble-dominated strike or the midrange-dominated strike.

Correct this relationship by adjusting rake to align the midrange with the tweeter. In the common case where the tweeter is mounted above the midrange driver, tilt the speaker back if the treble arrives first, and tilt the speaker forward if the midrange arrives first. Make this adjustment with a very small rotation of the rear spikes. Continue until the double strikes fuse tightly and land as one coherent event.

Phase XI: Azimuth—Last, and Only Last

At this point, the time has arrived to adjust the azimuth. Begin by listening to The Shins' "Sleeping Lessons." As you listen, use your mind's eye to visualize the "bubbles" rising and falling in space. Ideally, they should move straight up and straight down. In some cases, you may perceive the path of these bubbles leaning to the left, as illustrated in the drawing. (The examples shown below represents an extreme case.) Our goal is to stand them upright.

If the bubble path leans to the left, raise the inside of the speaker by **lowering the inside spikes** until the bubbles appear vertical. As you make this correction, you should also notice a reduction in the bloom and haze at the very top of the presentation. This reduction in "boominess" is an important confirmation that the adjustment is working.

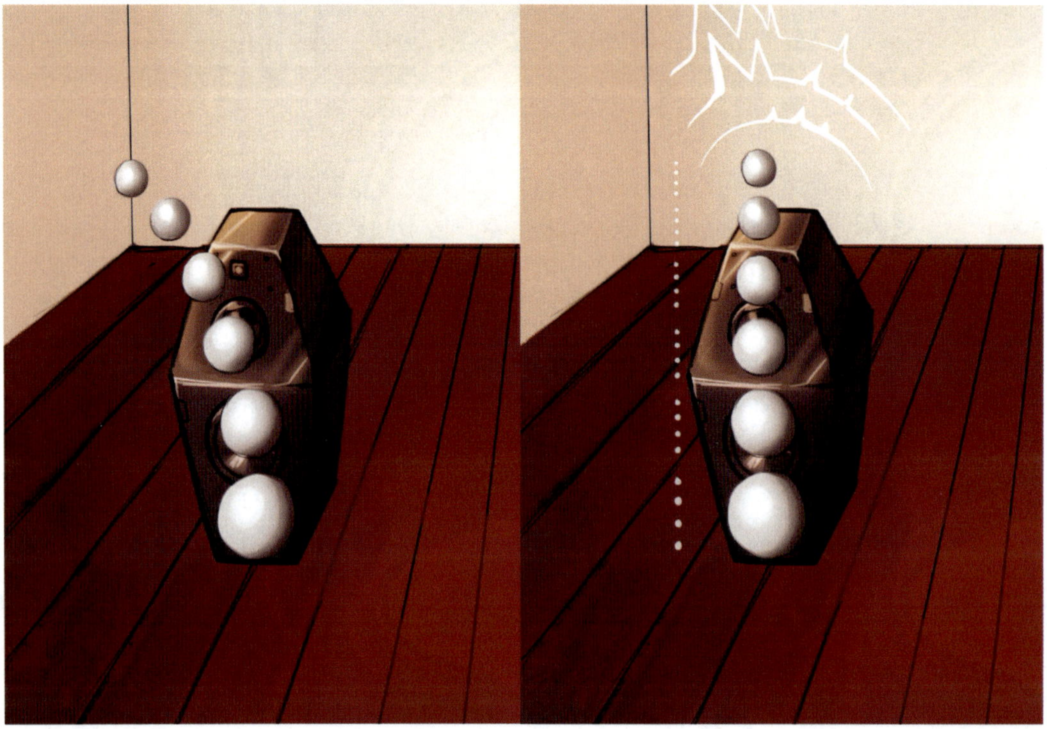

Correcting the azimuth reduces overlap between the tweeter and midrange grids and brings those grids into better alignment with the room. Because this adjustment directly affects tweeter-to-midrange timing, return to Kat Edmonson's "Lucky" and recheck the xylophone strikes. If necessary, make a very small rake adjustment to restore proper timing between those elements.

Once the bubbles appear vertical, the azimuth can be fine-tuned further. Listen to the first ten seconds of "Duende." The solo acoustic guitar should sound extremely precise and well defined. If you hear note slurring—especially as the guitar moves into higher frequencies—that indicates a small azimuth correction is still needed. If the guitar seems to lean outward, raise the inside spikes slightly. If it leans inward, raise the outside spikes instead. These adjustments should be made with very small rotations of the wrench, just enough to allow the threads to turn slightly. After each change, listen again and note whether the guitar becomes more articulate and stable.

Phase XII: Final Driver Integration with Rossini and Completion Signposts

For the final step in setting the left speaker, use *Rossini's L'italiana in Algeri* to fine-tune **driver-to-driver integration**. In the opening of *L'italiana in Algeri*, all of the instruments we hear—violin, cello, and bass—are playing pizzicato, with each string plucked rather than bowed. In the Orpheus Chamber Orchestra recording, this opening passage is performed very slowly, at approximately 52 beats per minute, and it should sound quiet, restrained, and delicate.

The first five measures of the score, shown in the following depiction, consist of a repeated pattern in which the strings play a group of eighth notes followed by an eighth rest. This structure creates a clearly defined transient attack with each pluck, followed by a long decay and a moment of silence before the next note. When the speaker is properly aligned, these elements—attack, decay, and silence—are all easy to hear and clearly separated.

Rossini's L'italiana in Algeri

Listen closely and identify each instrument individually. You should be able to hear the violin, cello, and bass as distinct sources, which allows you to evaluate the treble, midrange, and bass independently. In the ideal case, you hear an initial transient, a smooth decay, and then a clear gap representing the eighth rest before the next pluck. If the bass is shifted later in time, indicating that the bass pluck is arriving late relative to the others, raise the speaker a small amount until the bass and cello are playing together. This adjustment brings the bass driver's output forward in time so that its transient aligns more closely with the midrange and treble. Similarly, if you hear the violin is ahead of the cello then rake the speaker back a very small amount. As a more advanced refinement, listen not only for the bass pluck itself but also for the subsequent swell created by the resonance of the instrument's body. The correct order is for the pluck—the transient attack—to occur first, followed by the body bloom. If you hear the body swell arriving too early, that suggests excess bass bloom and indicates that a small height adjustment is needed to reduce it. Once the temporal order of the bass is correct, you can proceed with final alignment among the bass, midrange, and treble, confident that the drivers are now working together as a coherent whole.

Wrapping Up the Left Speaker

At this stage you may be wondering how to know when the left speaker is truly "finished"—what specific aural cues tell you it's time to stop adjusting and move on to the right speaker. The answer is that there are clear sonic signposts that indicate the left speaker has reached a stable, coherent state and is ready to serve as the reference for the right speaker.

One of the most surprising aspects of this setup method is how organized the sound can become even with the right speaker pushed back and turned away. A properly dialed-in left speaker will still produce a remarkably coherent soundstage and convincing imaging under these conditions. The best place to confirm this is *Shostakovich's Symphony No. 1*. It should sound exceptionally organized despite the right speaker being out of place. In the opening seconds, the hall should "come up" and stretch out before you, extending wall to wall and floor to ceiling. The trumpet should be solidly planted in the far-right rear of the soundstage. The flutes and oboes should sound pleasant and distinct rather than shrill. The clarinet should have a woody, natural character. Around 56 seconds, you will hear two bass drum hits: the second must be clearly lower in pitch than the first. Later, from roughly 2:34 to 2:40, there is a very quiet exchange where bass, cellos, and violins bounce back and forth. In this passage, you should hear a distinct silence after the bass and cello plucks, before the violin pluck arrives—silence that is obvious rather than implied.

Several other recordings provide equally useful confirmation that the left speaker is coherent and stable. In The Shins' "Sleeping Lessons," the bloom should be controlled, and the bubbles should be clearly defined rather than smeared together. In *Beethoven's Piano Concerto No. 1*, the piano should sound clear, twinkly, and level. In Kat Edmonson's "Lucky," the opening "hoooo'ing" should be controlled and precise rather than diffuse or grainy. In Yello's "Till Tomorrow," the trumpet should enter on the right at roughly 23 seconds and move in even steps across the soundstage. In "Cyril Davies," the saxophone should stand clearly on the right side of the soundstage with convincing placement.

All of these cues point to the same underlying quality: coherence and organization. The images may not yet be as precisely scaled or as perfectly rendered as they will be once the right speaker is fully matched—and that is fine. At this stage, what matters is that the left speaker is stable, musically convincing, and free of obvious defects.

This is also the moment to internalize a critical principle of the method. **The system will never "sound" better than it does with the left speaker dialed in while the right speaker is turned away.** That may seem counterintuitive, but it is central to the process. Our goal is not to "improve" the sound by adding the right speaker. The goal is to match the right speaker to the left without compromising what has been achieved. If you try to force the system to sound "better" once the right speaker is brought into position, you will almost certainly create a compromise. It is worth stating plainly and repeating: the sound will never be better than what you have with just the left speaker.

For that reason, before moving on to the right channel, remove any remaining defects you perceive in the left speaker's sound. If there is anything you are not completely satisfied with, you must identify what needs to be adjusted to correct it. In practice, this means listening carefully to the setup tracks—and, when needed, expanding to additional music—until you find the clue that reveals which of the six positional axes requires attention. If a specific defect seems impossible to eliminate, the next question is not "What move am I missing?" but "How accurate are these speakers?" Once positional accuracy is fully dialed in, the speaker's fundamental character will be revealed. If a driver rings at certain frequencies, those notes will always carry some degree of resonance. If the cabinet is not inert, the bass will always exhibit some bloom. Speaker placement is not a magic potion that can cure a poorly designed loudspeaker; it can only make the speaker perform at its highest potential.

When you are satisfied that the left speaker is performing at that potential, carefully tighten the locking rings on the spikes, and then proceed to bringing the right speaker into alignment with what you have established.

Evaluation of Changes (Optional, After Completion)

Once the left speaker is fully optimized, it becomes an ideal platform for evaluating system changes such as footers, shelves, cables, fuses, grounding devices, and other noise-reduction tweaks. With the right speaker turned away, the stereo illusion is not distracting you, and changes in noise floor become easier to hear. As the noise floor drops, the system becomes quieter, more transparent, more dynamic, and timing becomes easier to assess.

If you evaluate tweaks here, the three most revealing tools are Lucky for mechanical noise and vocal precision, Beethoven for note spacing and added "air" between piano events, and Rossini for silence, dynamic swell, and delicacy between pizzicato notes. Expect that meaningful noise reductions may require minor positional touch-ups. That is normal. Lower noise reveals finer grid lines.

Summary: Establishing the Reference Loudspeaker — Crossing Into Stereo

In Chapter 8, we moved beyond the initial neutral setup and focused on establishing the true working position of the left speaker. Through careful listening and small physical adjustments, we progressed from a provisional placement to a position defined by timing, tonal balance, and spatial stability.

This process unfolded in three stages. We began with a neutral placement that allowed the speaker to interact naturally with the room. From there, we refined the position by listening for improvements in clarity, tonal balance, and spatial focus. Finally, through subtle timing adjustments, we arrived at a position where the speaker revealed a stable and coherent acoustic presentation.

At this point, the left speaker is no longer simply "placed" in the room—it has become a reference loudspeaker. Its position now defines the timing and spatial framework that the rest of the system must follow.

This distinction is crucial. Up to this stage, the setup process has been exploratory, training your ears to recognize how small positional changes influence timing, tone, and spatial coherence. By aligning a single loudspeaker correctly, we have established a reliable acoustic reference—one that reveals the behavior of the room and the underlying structure of the recording with clarity.

This work may have felt slow and deliberate, but it serves a critical purpose. Without a reference point, stereo setup becomes guesswork. With one properly aligned loudspeaker, however, the system now has a stable foundation from which true stereo geometry can be built.

The next step is not simply to place the right speaker in a symmetrical position. Instead, the task is to bring the second speaker into alignment with the reference you have created, preserving everything that has been achieved with the left speaker while establishing the precise timing relationship required for stereo coherence.

This transition marks an important moment in the setup process. Up to now, you have been working primarily in mono, learning how a single loudspeaker interacts with the room and how timing reveals itself through tone and spatial cues. The next phase introduces the second speaker and begins the transformation from a single-speaker reference into a fully coherent stereo system.

Chapter 9 begins that process. There we will carefully introduce the right speaker and learn how to integrate it with the established reference so that both speakers work together as a single acoustic instrument.

When this alignment succeeds, the system will no longer sound like two independent sources. Instead, it will begin to reveal the organized spatial relationships embedded within the recording itself.

CHAPTER 9

ACHIEVING STEREO COHERENCE

Right Speaker Positioning Review

If you haven't yet completed the right-speaker exercises, stop here and go to Chapter Seven to review "Right Speaker Training and Exercises." Work through every exercise in that chapter before continuing.

And if you've already aligned the left speaker as detailed in Chapter Eight, **do not change anything about the left speaker** while you're doing those right-speaker exercises.

Preliminaries

Whew—the left channel is finished at last. Before we turn our attention to the right speaker, let us pause and recall its true purpose: to bring focus and order to the sound, shaping it both in space and in time. The aim is simple, yet profound—to cause the two loudspeakers to behave not as separate devices, but as a single, seamless instrument.

Once the left speaker is properly aligned with the room, the task before us is to bring the right into accord with it. When the two are in harmony, the soundstage opens and flourishes—broad, tall, and enveloping—revealing the full depth and dimension contained within the recording. Many listeners chase soundstage and stereo imaging as ends in themselves, yet these qualities arise naturally when the right speaker is properly matched to its mate. Instruments and voices then settle into precise, stable positions within the soundscape, as though etched delicately into the air.

Yet there is a deeper prize still: perfect timing. When sound waves across the entire frequency range arrive at your ears together from both speakers, something remarkable occurs. Figure 21 illustrates this moment. With true spatial and temporal alignment, the system becomes far greater than the sum of its parts. The noise floor appears to dissolve, leaving a sense of calm and stillness. From this quiet emerges extraordinary clarity and transparency—music not merely heard, but felt, resonating through both body and soul.

Figure 21: True Spatial and Temporal Alignment

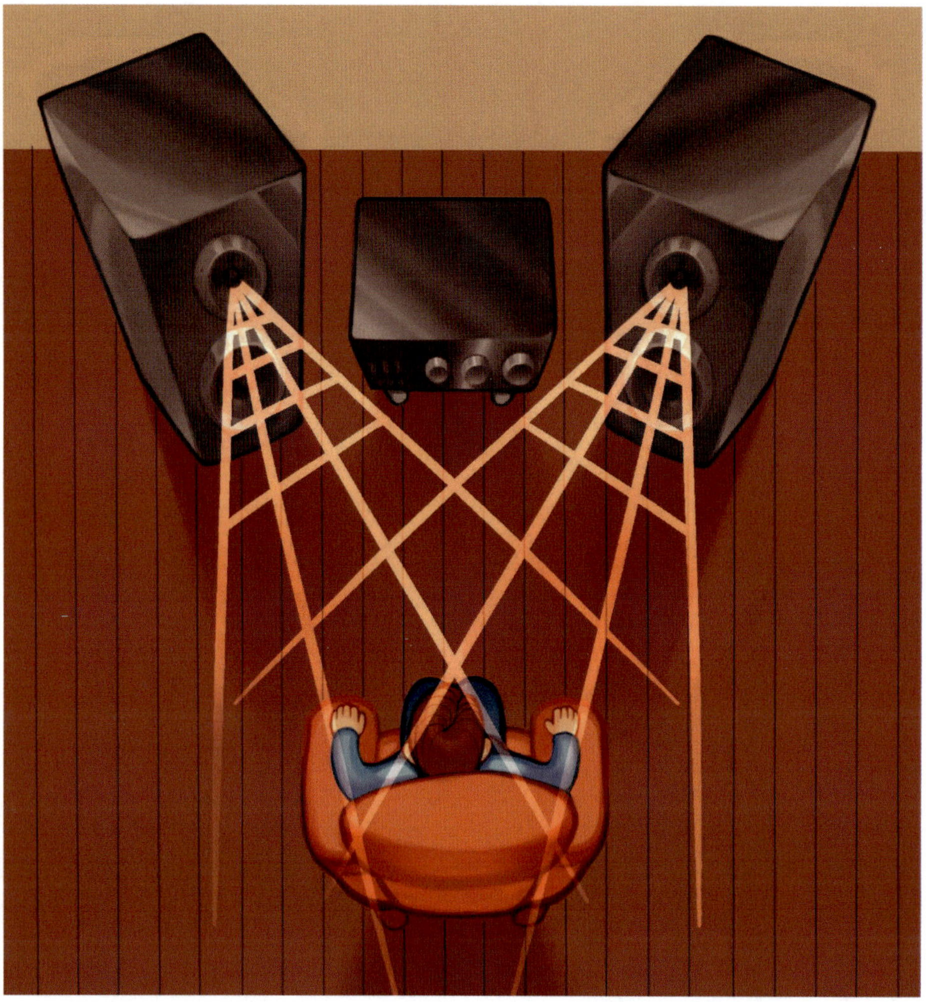

Earlier, I mentioned that the sound will never be better than what we achieve by listening to only the left speaker. Why is this, if we get all of the good stuff that comes along with aligning the right speaker? Our goal is simple: align the right speaker to match the tone and timing of the left speaker across the entire frequency spectrum. Unfortunately, there's a catch. Consider what makes up a stereo system—amplifiers, pre-amps, DACs, speakers, cables—each with separate left and right channels. Every component uses parts with manufacturing tolerances, typically around ±5%, or in more conscientious brands, ±1%. Because of these variations, the electrical properties of the left and right channels will never be exactly the same. These small discrepancies add up, making it more challenging to perfectly match the right speaker to the left than it should be. So, how is it that the sound is always better with the left speaker? Because any deviation from a perfect sonic match between the right and left speakers degrades the sound quality.

Enough dwelling on the challenges of stereo; it's time to move forward. We must work with what we have. The task of matching the right speaker to its mate is broken up into three large categories. First, we will utilize tools to measure the starting position of the right speaker. Second, we will utilize our music tools to organize the sound field in space. And third, we will use different music tools to organize the music temporally.

Just like with the left speaker, let's get a big-picture overview of the steps we will take with the right speaker before proceeding to the details of each step.

1. Transfer the measurements of the left speaker over to the right, resulting in a mirror image.
2. Make an overall assessment of the sound. Map out what is wrong.
3. Get the rake angle correct using "Till Tomorrow."
4. Feel for even pressure from left to right to make sure the right speaker is in the correct front-to-back position.
5. Get the front-to-back position close using "Poem of Chinese Drum."
6. Check lateral spacing — *Shostakovich's First Symphony.*
7. Refine the lateral spacing with Shostakovich and Beethoven's piano.
8. Get the toe-in nailed down using "Le Mal de Vivre" And "Till Tomorrow"
9. Micro-adjust for accurate timing using *L'Italiana in Algeri*, Beethoven, and Shostakovich.
10. Clean up the piano with small azimuth adjustments.
11. Make necessary compensating moves as you go along the path.

Although this process is laid out step by step, expect a fair amount of back and forth. In most cases, improving one aspect of the sound will increase clarity while also exposing subtle misalignments elsewhere (see the section on compensating moves). When you notice something that needs correction in any direction, make the appropriate adjustment—but be careful not to overcorrect in one area without checking the others. It's easy to chase a single fix so far that, once it's resolved, other elements end up out of alignment and become harder to sort out. A better approach is to make small, incremental adjustments, then reassess and apply compensating moves as needed. This keeps the overall balance intact throughout the process.

Proceed with caution—the right speaker can be deceptive. Many adjustments do not produce entirely independent effects. For example, pressure can be influenced by front-to-back movement, toe-in, rake, and lateral movement. Centering is affected by forward and lateral movement, rake, toe-in, and azimuth. The challenge arises when you follow a series of adjustments that seem to work but later realize another aspect of the sound is off. This means you've gone down the wrong path and will need to backtrack to find the correct setup. Reflecting back to the grid model, this means that one or more of the degrees of freedom is in the wrong grid space.

If you reach a point where the sound won't improve, first verify that your speaker measurements are correct. If distances and angles have changed significantly from your starting point, it may be best to reset the right speaker to its original "home" position and start over. This will likely be less effort than trying to work your way out of a setup mistake.

The Three-Stage Roadmap

The process of integrating the right speaker to achieve stereo coherence unfolds in three deliberate stages, each building on the one before it. We begin by establishing a precise physical reference, then organize the sound in space, and finally refine the system in time. Taken together, these stages move the right speaker from a measured starting position to a fully integrated partner of the left, allowing both speakers to operate as a single, coherent source.

Stage 1: Measure & Mirror

As the saying goes, "An ounce of prevention is worth a pound of cure." The best way to avoid mistakes and frustration is to position the right speaker as precisely as possible from the start.

What does "precise" mean? Ideally:

- Front-to-back and lateral positions within ±1/32 inch (0.03 inch)
- Height within 0.002 inches
- Rake and azimuth matched to 0.001 degrees
- Toe-in as precisely as you can visually see or measure

By starting with close-to-mirrored alignment, the need to find our way to the starting position by trial and error is eliminated. This ensures a cleaner, more accurate setup from the beginning.

Measuring the Left Speaker

Obtain all the measurements from the left speaker, being very careful not to bump it while getting this information.

- Use a tape measure to determine the distance from the centerline to the front inside spike.
- Measure the distance from the front wall to both front (or rear) spikes.
- Measure the height from the floor to the bottom of the front of the speaker at the midpoint between the spikes.
- Use the Digipas level to record the rake and azimuth of the speaker.

Place a piece of painter's tape on the floor between the centerline and the left speaker. As you are gathering the numbers, write them on the painter's tape with a Sharpie. Once you have this information, transfer it to the right speaker.

Before moving the right speaker into position, a discussion about issues with the front wall is needed. Measuring the distance from the wall behind the speaker with a tape measure is dubious. No wall is perfectly square. The illustration below is exaggerated, making the point clearer. If we measure from the wall behind the speaker (the front wall) in this room, there would be a large discrepancy between that measurement to the center of the front spike and the actual distance required to have the speakers sitting in the same horizontal plane. In fact, when comparing the distance from the left and right speakers to the front wall, a ¼" difference in distance is commonly found.

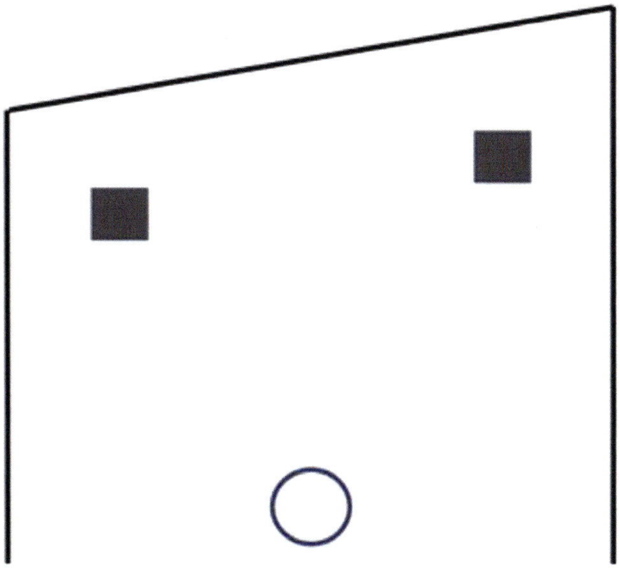

The relative importance of two possibilities must be considered: the speaker's position in the room or the position of the speaker relative to its mate. The room is undoubtedly strong. It has pressure modes, and the speaker's position relative to the wall will interact with them. However, it bears out from experience that the speaker's position relative to its mate is a far more important consideration. The sound is much clearer and more dynamic when the speakers are in the same horizontal plane. This should not be a big surprise, since having the correct phase relationship between the two channels across all frequencies is a fundamental concept in stereophonic sound. This can only be achieved if the speakers are in the same horizontal plane.

As seen in the picture below, the issue with non-square walls can be circumvented using a 12- or 16-beam laser level like the one pictured. The laser beam on this type of level is offset to the side of the unit. This enables the placement of one beam along the centerline of the room. The perpendicular beam can then be positioned to intersect a reference point on the left speaker, such as the inside corner or the center of the front spike. The right speaker can be placed so that the beam intersects the same location as on the left speaker.

The room's centerline must be marked as accurately as possible. Follow the instructions outlined in Chapter 6. You will need two points to align the laser with the centerline. Choose two points about 18 inches apart. Mark the center on a piece of blue painter's tape using an ultra-fine-line Sharpie marker. Be precise. Make the marker dot as small as possible—smaller than, or the same size as, the laser line's thickness.

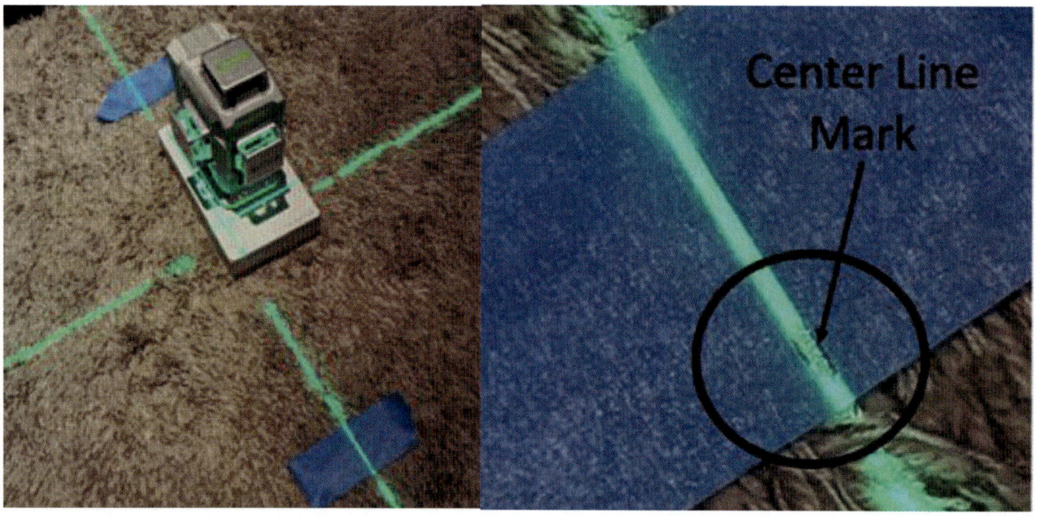

This next real-world example illustrates the significance of wall variance. In a recent setup, the left speaker measured 35 inches from the wall to the center of the front inside spike. You can see the laser forming a line through the middle of the left speaker's spike threads. Using a tape measure, the laser line's center is just under 35 inches from the wall. In the images of the right speaker, note that the tape measure also shows 35 inches to the center of the spike thread. You'll notice the laser crosses the tape measure at 35 3/16 inches—clearly in front of the spike. The speaker would be 3/16 inch too far back if positioned using the measurement from the front wall. This places it several grid spaces back and will cause significant issues when trying to align the right speaker.

Left Speaker

Right Speaker

Right Speaker

Right Speaker

Moving the Right Speaker Into Position

With the positional measurements of the left speaker in hand—and a reliable method for reproducing them—we are ready to turn the right speaker back toward the listening position. Before doing so, mark the exact spot on the floor where the front-inside foot of the right speaker will land.

Align the laser as precisely as possible. Place a tape measure on the floor and extend it toward the right speaker. Position the tape measure so that the laser line crosses it at the same distance used for the left speaker. At the end of the tape measure, place a piece of painter's tape on the floor. Using a marker, draw a line on the tape at the point where the laser crosses. This mark in the next picture defines the exact target location for positioning the right speaker.

Turn the right speaker and move it to the "home" position so that the front-inside foot is touching the tape. Visually approximate the toe-in by sighting down the left speaker's side from the listening position and matching the right speaker to what you see on the left. When adjusting the toe-in, ensure the front-inside foot stays on the mark.

Raise the front of the speaker to the correct height. Then adjust the rake using the rear spikes and the azimuth using both the inside and outside spikes so that the speaker pivots on the centerline. (This will help maintain the proper height.) Recheck the height and adjust these settings until the speaker matches the height, rake, and azimuth.

Next, set up the 16-line laser and line up the right speaker with the left. You will need to go through this process two or three times until the speaker is fully matched. Take your time. Don't rush. Expect this to take an hour or even longer the first time you go through the process. The time you spend here will be well worth it. If you need to take a five-minute break to catch your breath, stretch your back, and vent some frustration, that is fine. Once you are done, check the speaker connections to ensure they are tight, since we have been moving this speaker around.

Pro Tip: *For setting the height of the right speaker, it is easier to use a "gauge" of some kind. You could cut a wooden block that barely fits under the left speaker, or you could use something threaded that fits under the left speaker. Then, using this, you raise the right speaker until the "gauge" barely fits. (See picture following picture.)*

Toe-in is the most complex measurement to transfer to the right speaker. There are a few methods you can try, but in practice, it remains the least accurate initial setting. You can experiment with these or feel free to invent your own method that works for your specific speaker. Once the process of massaging the speaker into position has started, I prefer to toe in the speaker. It is easier for me to hear, and it is easier to execute because I can reach the inside corner more easily. So, when setting the initial toe-in, aim for it to be as close as possible. If you're unsure, I recommend leaving the speaker slightly toed-out.

Two Ways to Set Initial Toe-in

First, you can sit in the listening position and use your eyes to sight down the left speaker. After you have established a visual marker—such as the amount of the sidewall you can see—you can set the toe-in of the right speaker so that it visually matches.

Second, you can use the laser to help with alignment. We've already described the benefits of using the multi-axis laser when matching the right speaker's position to the established left speaker. Be mindful that a significant discrepancy (1mm) in manufacturing tolerance can really throw things off. For example, if you are using the spike as a reference point but the absolute position of the threaded spike from the front baffle is not matched then the speakers will not be matched either. The important thing is that when finished, the right speaker is a mirror image of the left speaker relative to the system centerline and the listening position.

We can extend the mirror image concept to set toe-in by creating a visual map on each speaker using the multi-axis laser to compare how they sit in space. Position the multi-axis laser so that the lateral laser line hits the left speaker somewhere in the left half of the front baffle. (The front-to-back line must remain on the center line of the room.) This creates radius lines on the driver cones and straight lines on other parts of the baffle — together forming a visual picture that can be matched on the right speaker. As you rotate the right speaker, you match how the two speakers reflect the laser as seen from the listening position.

These images show a real-world example of a left and right speaker. You can see that the laser's image on the right matches very closely to the one on the left. Note the highlighted area. You can see a small light reflection between the driver's surround material and the baffle ring. If the right speaker is moved less than 1mm that small piece of light reflection will disappear. When using this method, it is those small details that will show you how closely matched the two speakers are.

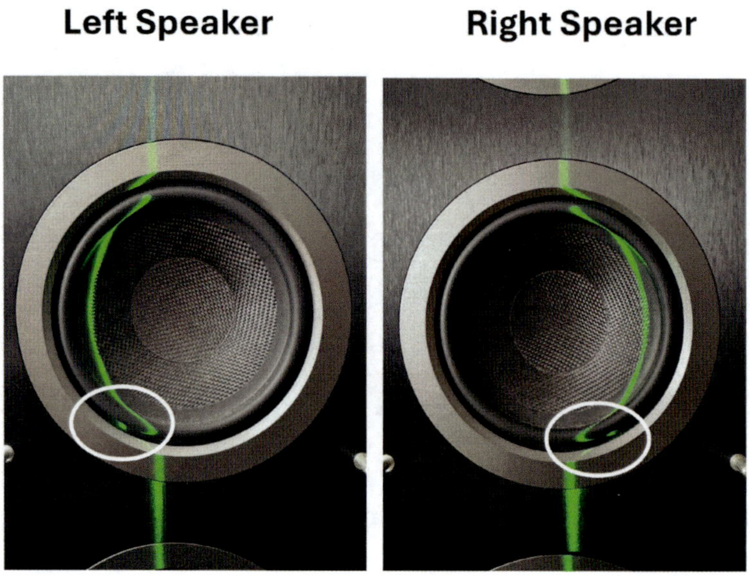

After rotating the speaker, verify that the inside front footer remains in the same position you placed it to match the left speaker. If that position changed during the rotating process, you would need to reposition the footer and check the reflection again. This is inherently iterative —each adjustment in one axis shifts the others, so expect to cycle through positioning and verification several times before everything converges. Depending on how exacting you want to be, this can take longer than you might expect. However, if the laser is very well matched, then you will likely find the first round of listening tests will already be correct.

I find that measuring from the speaker's side wall to a laser line gives better results than relying on your eyes alone to match reflections. Bear in mind, this method of matching right to left considers all six axes of movement. If you are having trouble getting the reflections to look or measure the same, there may be more than one positional change to consider. On speaker cabinets with a radiused corner and no well-defined place from which to measure, try measuring from the edge of a driver. If that's not feasible either, you will have to use your eyes to discern the difference in the two reflections.

For example, if a laser line falls 1¼ inch; from the outside of the left speaker cabinet but only 1 inch from the outside of the right, you could equalize that difference through toe-in rotation, front-to-back position, or lateral position. If the differential is very small and all other reflections look spot-on, the discrepancy could even be speaker azimuth. But remember, for this initial setup we are attempting to mirror-image the right speaker to the left, so matching speaker height, rake angle, and azimuth precisely matters just as much as lateral and front-to-back positioning. This multi-axis laser technique is a good illustration of how every positional change you make in one axis affects all the other axes.

Sound Check

Now that all the measurements have been transferred to the right speaker, it is time for a sound check. I suggest starting with a vocal track, such as Leonard Cohen's "Treaty." In theory, the image of his voice should be solid and centered. Of course, the sound will not be perfect at this point, but it should be relatively centered, indicating that the right speaker is positioned correctly. If the image is substantially off-center (meaning pulled more than 25% of the distance to the left or right), recheck your measurements. If you recheck your measurements and everything is correct, but the image is still pulled to one side, it could be due to an asymmetry in the room or some other factor. We will work to correct the sound. If Leonard is relatively centered, you can take this time to listen to a wide range of tracks we will use for setup and take some notes.

Stage 2: What "Spatial Organization" Means

Recognizing the Emerging Structure

The right speaker is in its "home" position, and we are ready to proceed with the alignment process that acoustically matches it to the left speaker. We want every instrument or singer to occupy its proper location, regardless of the pitch being played or sung. For example, most vocal tracks recorded in a studio have the singer perfectly centered. It should not matter whether we are listening to the baritone voice of Leonard Cohen, the alto voice of Annie Lennox, or the soprano voice of Diana Damrau; they should all be centered. We will use several musical tools that highlight specific frequency ranges to ask a simple question relative to the illustration below: *Is the image centered?* If not, we need to adjust the speaker to correct it.

Beyond overall centering, other imperfections in the sound field must be addressed. You may notice that I've selected several electronic music tracks for tuning the right speaker. Whether or not you enjoy electronic music is irrelevant—if anything, it's better if you don't. That way, you won't be distracted by the music itself and can focus on identifying and correcting sonic defects. Reference tracks should be treated purely as test tones. If you find yourself getting drawn into the music, that's actually a good sign—it means you're close to achieving a proper setup. Still, try to remain focused on the process.

Electronic music is ideal for the initial phase of fine-tuning the right channel because it is perfectly timed—there is no variation from musicians playing slightly off-beat. It is also highly symmetrical, with the left and right channels often being identical copies. This means that whatever we hear on the left should be precisely mirrored on the right. With this framework in mind, we now shift from understanding what a correctly organized sound field should feel like to diagnosing what is actually happening in your own system.

Interpreting and Acting on What You Hear

Figures 22 and 23 illustrate the kinds of musical impressions that may be evoked when playing electronic music. They depict a trumpet, a shaker, and three clicks or pops. In the first image, the sounds on the right side are distorted and displaced relative to those on the left. If this matches what you are hearing, it provides clear guidance on how the right speaker should be adjusted.

Begin by noting that all of the images on the right appear low in height. This indicates that the right speaker needs to be raked back in order to raise the image. As the image height is corrected, it is common for the image to shift laterally as well. Next, observe that the images on the right appear too far forward relative to those on the left. This suggests that the right speaker should be moved backward. Finally, notice that the images on the right are blurred and lacking focus. This indicates that additional toe-in is required for the right speaker.

Once the front-to-back position, rake, and toe-in have been adjusted, a new set of images will emerge on the right. At that point, you can reassess the sound field and determine which direction the speaker needs to move next.

Figure 22: Shifted and Distorted Images

Figure 23: Perfectly Symmetrical Images

Initial Assessment of the Right Channel

Like the prior illustrations, we need an overall assessment to understand what we are hearing incorrectly and to determine the direction the speaker should move. To do this, I suggest creating a map. I have included an example, along with space for you to draw what you observe in your own system.

More often than not, images are pulled to the left, regardless of the measured precision of the right speaker's placement. I believe it is plausible that this is related to the **Franssen effect**. This effect was discovered by Nico Franssen in 1960, and the research was further refined and published by W. B. Hartman and B. Rakerd in 1989 (see reference). The Franssen effect states that sound localization is at least partially determined by the location of a sound's initial transient. In our current situation, we have worked diligently to refine the transient response of the left speaker, while the right speaker is only loosely positioned. This imbalance creates a scenario in which the image is pulled to the left due to the sharper transient. As we begin improving the coherence of the right speaker, the image will naturally resolve toward the center.

We will assess four of the six degrees of freedom: rake, front-to-back, lateral, and toe-in. For the remaining two degrees of freedom, recall that the height of the right speaker will remain mostly unchanged and that azimuth is reserved for last as a final "touch-up."

Begin by playing Yello's "Till Tomorrow." This track provides a great deal of useful information. Start by asking a simple question: where are the trumpets on the left compared with those on the right? Listen to the trumpet as it moves from right to left starting at the 22-second mark. On the image below, draw a line representing the path of the trumpets as you hear them move across the soundstage.

Figure 24: Oops. This Path is too High on the Left

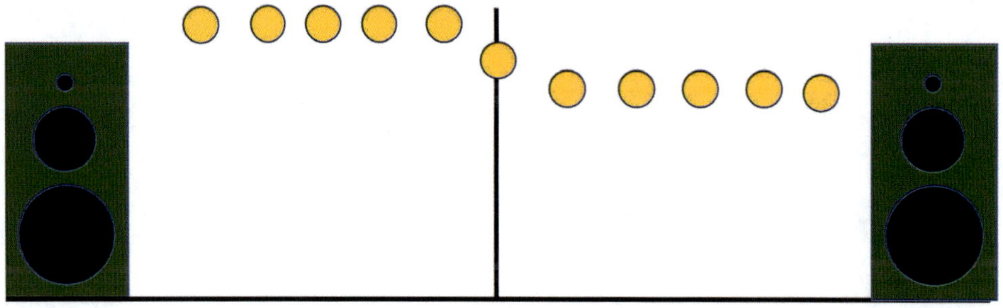

Figure 25: What Does Your Path Look Like?

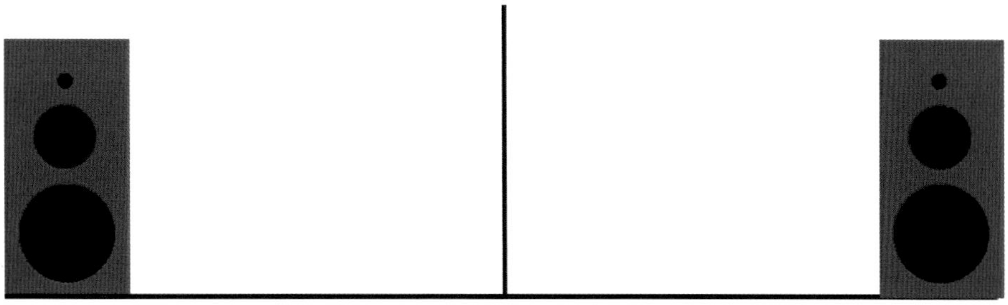

Next, listen to the opening section of Yello's "Till Tomorrow" again. You will hear the bass beat, but more importantly, you should *feel* the bass pulses. Focus carefully on where you feel the pressure on your body. (If you do not feel any bass pressure at all, something has gone wrong with the setup of the left speaker.)

Play the intro as many times as necessary until you can clearly identify where—on your torso and head—you feel pressure, and where you do not. Shade the corresponding area on the image below to reflect what you are experiencing. I have shaded the example image on the right based on a recent system setup to illustrate what this might look like. In that example, the pressure was much stronger on the left side, which is not correct. The bass pressure should feel even and centered across the torso.

Figure 26: Oops. Bass Pressure Not Centered

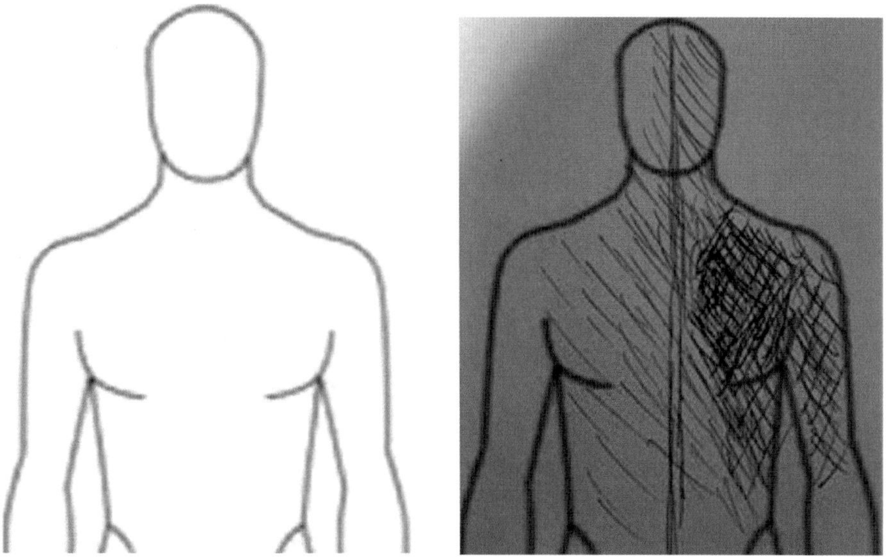

The final section of "Till Tomorrow" to examine begins at roughly the one-minute mark and continues to about 1:20. During this passage, the background bass beat remains, while a single trumpet appears centered between the speakers. The trumpet has a distinctly *breathy* quality. Its notes should be mostly contained within a circular image, positioned squarely at the center of the soundstage.

Listen carefully and ask yourself: does the trumpet's breathiness stream to the right, or does it stream to the left, or is it centered? On the diagram provided, mark how you perceive the trumpet's sound and the direction in which that breathiness seems to flow.

Figure 27: The Trumpet is Pulled Left and Streaming to the Left

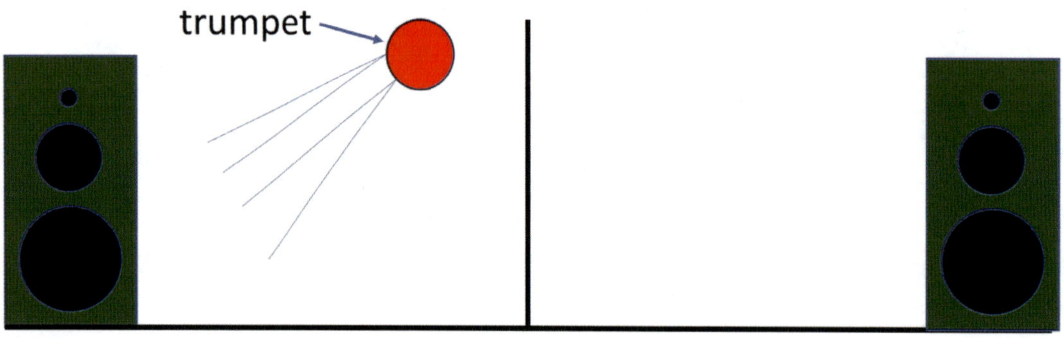

Figure 28: How Does the Trumpet Sound?

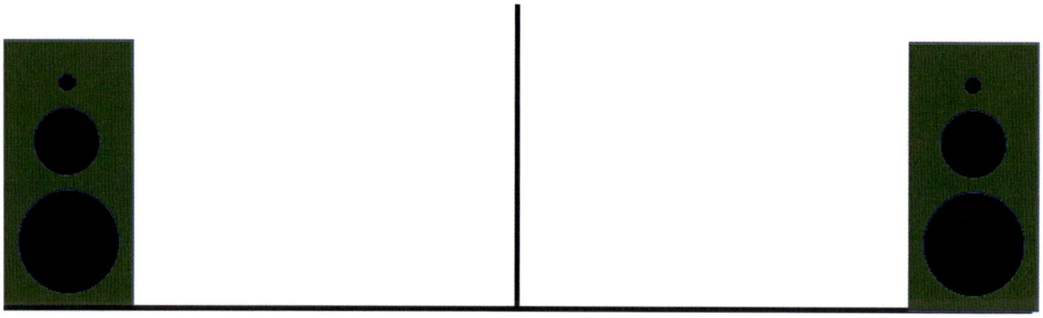

Next, listen to "Poem of Chinese Drum" and "Treaty." On the diagram, draw a circle to indicate the position of the drum and its perceived size. Then do the same for Leonard Cohen's voice, noting both its location and scale within the soundstage.

Figure 29: Oops. Bass Images Pull Right

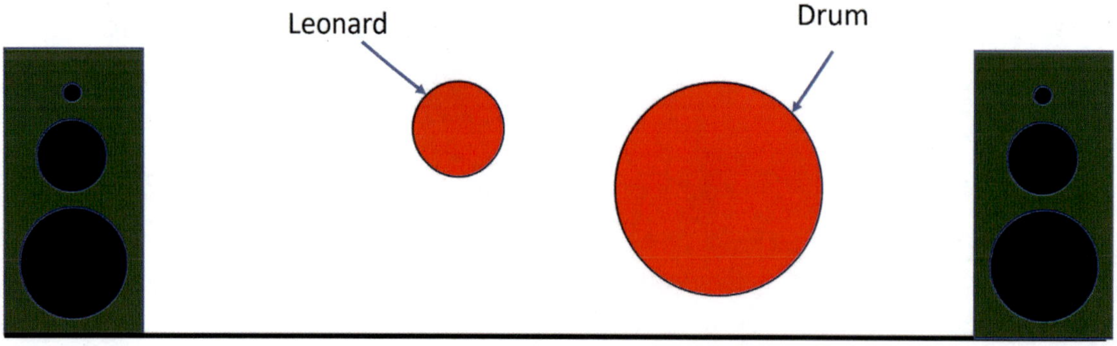

Figure 30: Where are Your Bass Images?

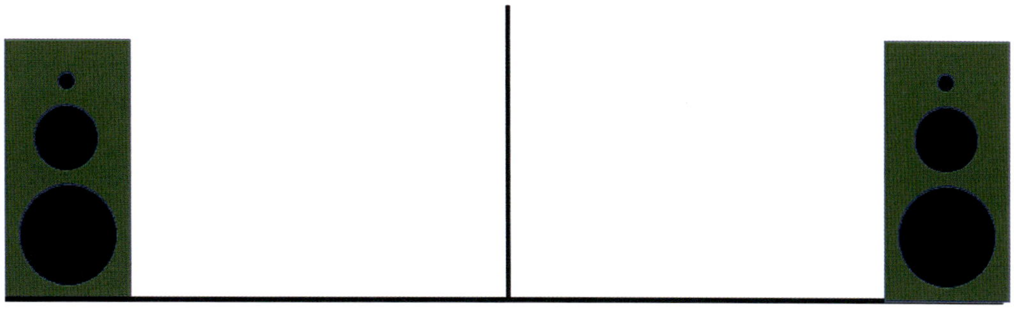

Listen to "Dancin' in the Dark" and "La Mar De Vivre." On the diagram, draw a circle representing the location of Ruth's and Cecile's voices within the soundstage.

Figure 31: Oops. Midrange Images Pulled Left

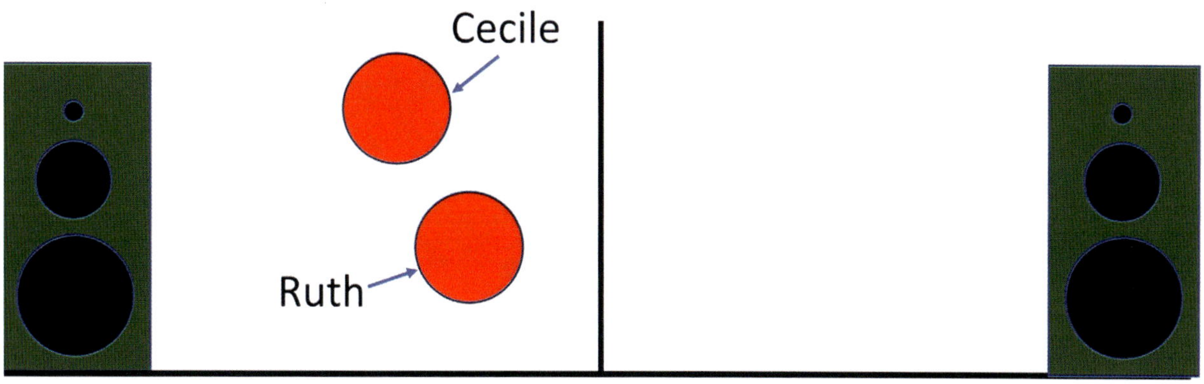

Figure 32: Where are Your Midrange Images?

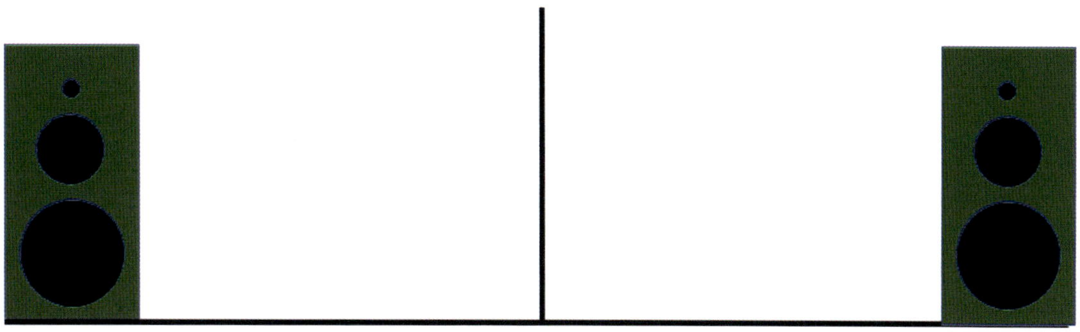

The final element of the map is lateral spacing. Listen to the Allegretto of *Shostakovich's Symphony No. 1* between 29 and 32 seconds. Do you hear the orchestra evenly spread across the soundstage, or do you hear two pillars of sound clustered near the speakers? Because we are measuring distance from the centerline, this dimension is typically the most accurate and, therefore, most likely to sound correct. On the diagram, mark where you hear the bulk of the orchestra between 29 and 32 seconds.

Figure 33: "Pillars" of High Density Near the Speakers

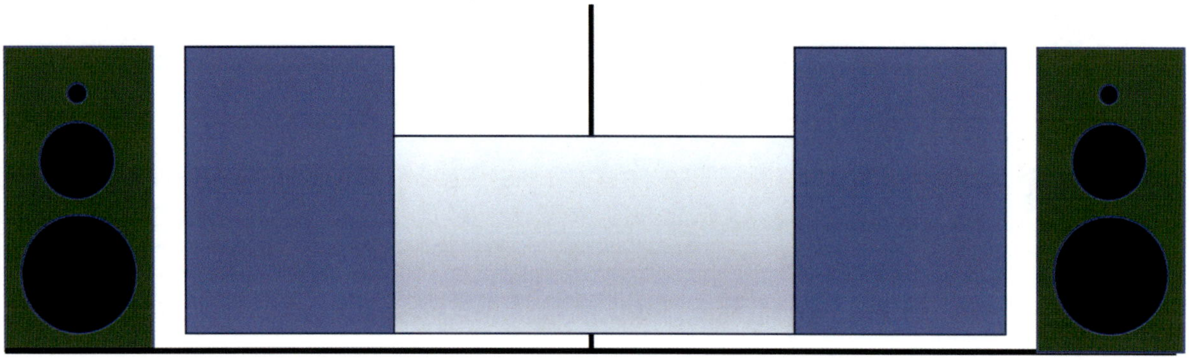

Figure 34: Where is the Orchestra?

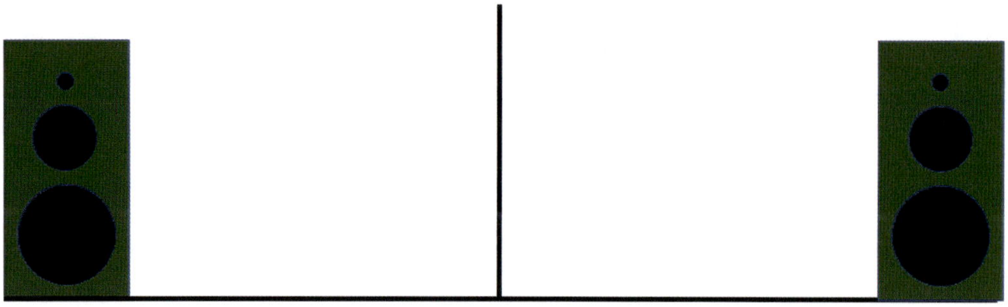

The observations and drawings you completed indicate the directions the right speaker needs to move to match the left.

- The moving trumpet in "Till Tomorrow" tells you how to adjust the rake angle.
- The big drum in "Poem of Chinese Drum" tells you whether the right speaker needs to move forward or backward.
- Leonard Cohen and Shostakovich tell you about lateral spacing.
- The centered trumpet in "Till Tomorrow," along with Ruth and Cecile, provides toe-in information.

Now that we have a map, we are ready to interpret it. Let's use the data I collected as an illustrative example. In my setup:

- The moving trumpet was high on the left — rake the right speaker back.
- The big drum was pulled right, with more pressure on the right — push the right speaker back.
- The single trumpet was breathy on the left — toe-in the right speaker.
- Shostakovich was separated in the center — move the right speaker laterally inward.

Before making your first adjustment to the right speaker, consider the implications of moving in each direction and how those moves affect the others. This was covered in the section on compensating moves. The following tables show the effects.

	Front/back	Lateral	Toe	Rake
Forward	N/A	No Effect	Toe-out	Forward
Inward	No Effect	N/A	Toe-in	No-effect
Toe-in	Forward	Inward	N/A	No-effect
Rake Forward	Forward	Inward	No Effect	N/A

	Front/Back	Lateral	Toe	Rake
Backward	N/A	No effect	Toe-in	Backward
Outward	No Effect	N/A	Toe-out	No effect
Toe-out	Backward	Outward	N/A	No-effect
Rake Backward	Backward	Outward	No effet	N/A

For example, if you determine that the right speaker needs both toe-in and a backward movement, pushing the speaker back will also help resolve part of the required toe-in. By contrast, adding toe-in will move the acoustic center forward and can worsen the need to move the speaker backward.

I want to emphasize that this process will involve frequent returns to adjustments that initially seemed "good enough." Think of the setup as moving toward a target through multiple rounds of refinement. To begin, we will adjust the rake, then make front-to-back movements, followed by toe-in adjustments and small lateral bumps. Consider this **Round 1**. We will then return and make smaller adjustments to each of these in **Round 2**, repeating the cycle until the right speaker matches the left and the results are satisfactory. This process is not strictly linear. While I can describe what to listen for and which adjustments typically address specific issues, I cannot provide a precise, move-by-move sequence. You must rely on your ears and continually compare the right speaker to the left and ask yourself: "are things where they belong?"

Recall that there are six possible adjustments for the right speaker: front-to-back, left-to-right, up-and-down, toe-in, rake, and azimuth. We must choose one to begin with, as it is highly likely that all of them, except height, will require some adjustment. We will not change the height of the right speaker's front edge. The speaker's front height relative to the floor will remain fixed. The floor is the closest boundary to the woofers, and both speakers rest on this surface, so we want their interaction with that boundary to be the same. Because most floors are not level, one speaker will sit lower in space than the other. We must determine which speaker is lower and adjust the rake accordingly to compensate for the height difference between the left and right speakers.

Round 1: Coarse Spatial Alignment

Rake First: Leveling Image Height

To begin work on the right speaker, we will start by adjusting the rake angle. You may wonder why we do not begin with the front-to-back position. The reason is that rake has a significant influence on how we perceive bass, image centering, and timing. Until the rake angle is close to correct, it is difficult—if not impossible—to accurately assess the front-to-back position. For this reason, rake should be addressed first.

The primary purpose of the rake angle is to match the apparent height of objects on the right side of the soundstage with those on the left. Because most floors are not level, we need an initial indication of whether the right speaker should be raked forward or backward. Begin by unpacking a self-leveling laser level. Place it on a tripod, a small table, or even the edge of a chair. As long as it is not excessively tilted, it will auto-level. (Ensure the locking switch is disengaged)

Observe where the laser beam crosses the left speaker, using a fixed reference point such as a driver bolt or another clearly defined feature. Then observe where the laser crosses the right speaker at the same reference point. If the laser hits higher on the right speaker, this indicates that the right speaker is physically lower in space and that the floor slopes downward toward it. To compensate for this height difference, the right speaker will typically need to rake backward. Conversely, if the laser hits lower on the right speaker—meaning the right speaker is physically higher—then the floor slopes upward toward it, and the right speaker will typically need to rake forward to compensate. The images on the following page show the laser crossing the right speaker about 1/8" higher than the left.

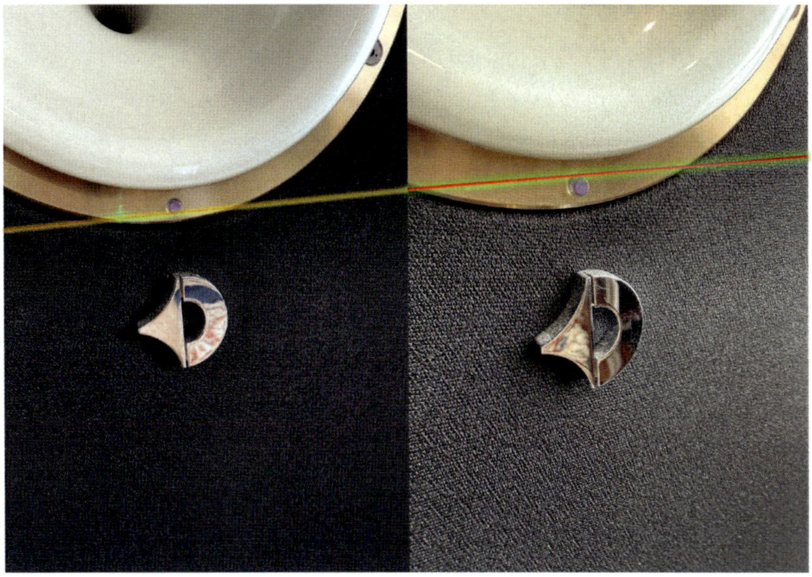

Yello's "Till Tomorrow" is an excellent musical tool for revealing left-to-right height differences in the soundstage. You have already listened to this track and recorded your observations on the map. At approximately the 22-second mark, you will hear a series of trumpet blasts that begin on the right and move steadily to the left. The trumpet blasts start softly, increase in volume as they approach the center, and then gradually soften again as they continue leftward. More importantly, all of the blasts are the same pitch and, from an imaging perspective, should move across the soundstage at a constant height.

At this stage of the setup, it is very common for the trumpet image to shift either upward or downward as it reaches the center point. Figure 35 depicts an example in which the trumpet shifts upward once it reaches the center of the soundstage.

Figure 35: Trumpet Shifts Upward at Center of Soundstage.

If the trumpet shifts upward as it moves to the left, this indicates that the soundstage is low on the right. Which direction we should initially rake the speaker depends on whether the right speaker is lower or higher in space relative to the left speaker. If the right speaker is lower in space, then we should rake the right speaker back until the trumpet is level. If the right speaker is higher in space, then we should rake the right speaker forward until the soundstage is level. Adjust the rake by rotating the rear spikes. Begin with approximately **1/16 of a turn** and listen to the effect this has on the trumpet's movement from right to left. Continue adjusting until the trumpet sounds level across the soundstage. Once the trumpet is level, revisit the other three observations in **"Till Tomorrow."** Did the bass pulse change? Did the pops and clicks move? Did the echo chamber change? We will not act directly on this information at this stage. The purpose of this reassessment is to reinforce the idea that adjusting the rake angle affects multiple aspects of the sound simultaneously.

Rake Zones Again?

Recall the concept of **"Rake Zones"** discussed in Chapter Five under the *"Timing and Alignment Zones"* subheading. This is an ideal moment to explore that concept further.

At this point, you have aligned one degree of freedom—the rake angle—to a grid line. Granted, this may not be the *correct* grid line. The complication arises from the floor. If your floor is very close to level, you are likely on the correct grid. The more out of level the floor is, the greater the chance that you are on the wrong grid. This may sound unclear—and that's understandable. The most effective way to grasp this concept is through direct experience.

Begin raking the right speaker backward in very small increments, approximately **0.002 to 0.003 degrees** at a time. With each adjustment, listen carefully and note the perceived height of the trumpet's image. As you continue raking the speaker backward, you should hear the trumpet rise progressively higher on the right side of the soundstage. Then, quite suddenly, the image will drop lower. Continue increasing the rake angle until the trumpet once again matches the height of the left side.

Record this new reading on the DigiPas level. By subtracting this value from your initial rake reading, you will obtain a close approximation of the angular distance between rake zones.

The rotational spacing between grid zones is system-dependent. I cannot specify how many degrees of rake angle separate one zone from the next in your system. However, you now have a concrete measurement of how much rotation it takes *in your setup* to move from one zone to another. This information will be extremely useful in the future if you need to deliberately move the speaker into a different rake grid.

Before continuing, return the speaker to the original rake angle that produced a level soundstage. Finally, confirm that the azimuth once again matches that of the left speaker. Note that you have changed the rake angle by adjusting only the rear spikes. Any necessary correction to the azimuth can be accomplished by adjusting only a rear spike.

Coarse Front-to-Back: Center the Drum

Having completed the rake adjustment, we now need to align the **front-to-back** position of the right speaker. Even if this position were perfectly aligned at the outset, changing the rake—tilting the speaker forward or backward—will have shifted the acoustic center forward or backward as well. At a minimum, the speaker's position must be adjusted to compensate for this change. In practice, absolute precision is difficult to achieve: the laser line has measurable width, and we are likely off by at least **1/64 of an inch**. Our goal is to bring the right speaker back into the same plane as the left.

Play "Poem of Chinese Drum" and ask yourself a simple question: *Is the big drum centered?* If the drum image is pulled to either the right or the left, the front-to-back position of the right speaker is significantly off.

- If the drum image is pulled to the **right**, bump the right speaker **backward**.
- If the drum image is pulled to the **left**, bump the right speaker **forward**.

Figure 36: Drum Centered

Figure 37: Drum Slightly Right

Figure 38: Drum Far Right

Continue making adjustments until the drum image is centered. The more the drum is displaced to one side, the larger the initial bumps should be. For example, if the drum is pulled more than one full drum width to either side, begin with firm, two-handed palm strikes to the front or back of the speaker. After each bump, listen carefully to determine how much the image shifts relative to the force applied. Once the drum image is centered, proceed to the next section.

First Compensating Move: Rake Adjustment

It is now time to make the first compensating move. Adjusting the speaker forward or backward alters the rake angle required to maintain a level soundstage. Listen to the trumpet once again and make the appropriate correction using the rear spikes to re-level the soundstage. If you bumped the speaker forward, you should hear that it now needs to be raked forward. If you bumped the speaker backward, you should hear that it needs to be raked backward.

Coarse Lateral Adjustment: Close the Gap

The lateral position of the right speaker is typically the most accurate starting point, since both speakers were measured from the same centerline reference. However, if the front-to-back position had to be moved significantly, the speaker may have drifted off that line.

We will use two tracks to assess lateral spacing. First, listen to Leonard Cohen's "Treaty" and note the location of his voice. Next, listen to the Allegretto from *Shostakovich's Symphony No. 1* between **29 and 33 seconds**. Ask yourself whether the orchestra sounds like two pillars of sound near the speakers with lower density in the center, or whether it is evenly spread across the soundstage. If the orchestra sounds evenly distributed, a finer tool will be needed to determine any remaining adjustment. If you hear lower density in the center, the right speaker needs to move laterally inward. Bump the speaker inward until the orchestra sounds evenly spaced.

Orchestra Crowded Near The Speakers

Now listen to Leonard's voice again. Is it centered, or is it still pulled off-center? If you moved the speaker inward to connect the orchestra, his voice is likely much closer to the center. We will revisit this in the next round of adjustments.

Next Compensating Move

Moving the speaker laterally does not, in itself, change the front-to-back position or the rake angle. Although, during the initial phase—when movements are relatively large—the speaker can easily drift off its intended line. Return to the trumpet and recheck the rake angle, adjusting it if necessary. Then verify the front-to-back position by listening for the drum's centering and correct it as needed. These compensating adjustments should be small.

Coarse Toe-In: Center Voices and Focus

Toe-in is used to center midrange images and sharpen focus. Begin by listening to Ruth Moody and Cecile McLorin Salvant. Their voices should sound perfectly centered. If they are not, note where each voice appears within the soundstage.

If their voices are pulled to the left, then toe-in is required for the right speaker. If the voices are pulled to the right, then toe-out is required for the right speaker. To accurately toe-in or toe-out the speaker, use one hand to secure the front inside spike position and then bump the rear corner of the speaker with the other hand. You will bump the inside corner to toe-in and the outside corner to toe-out. With each toe-in/toe-out bump, reassess Ruth and Cecile's position.

This illustration depicts singers drifting left with increasing frequency. The right speaker needs toe-in to correct this.

More Compensating Moves: Keep the Balance

Another round of compensating moves is now required. If the right speaker was toed in, the acoustic center moved forward and inward. If it was toed out, the acoustic center moved backward and outward. Corrections will be needed to the front-to-back position, rake angle, and lateral spacing.

As with the left speaker, I **strongly recommend** that you **do not** adjust the azimuth at this point. Leave it fixed at the same value used for the left speaker. The grid spacing for azimuth is extremely fine, and if you begin adjusting it now, you are likely to end up several grid lines off by the time the process is complete. Ensure it has the same value as the left speaker.

Round 2: Fine Spatial Refinement Fine-Tuning the Spatial Arrangement

We have now completed the first round of adjustments to rake, front-to-back, lateral position, and toe-in. The right speaker should now be a much closer acoustic match to the left than it was in its initial measured position.

Soundstage events should be more accurately placed, but there is still room for improvement. It is time to reassess the sound with a more critical ear and shape it with better tools.

Finer Rake: Confirm with ""Yellow Brick Road"

"Till Tomorrow" is an excellent track for getting close to a level soundstage, but other recordings reveal soundstage height even more clearly. Once the trumpet sounds relatively level, switch to Sara Bareilles' **"Yellow Brick Road."** The opening piano chords should sound level, as if the piano keys are stretched evenly between the speakers (see the illustrations in the Reference Songs chapter). If the trumpet already sounds level, the adjustments needed to level the piano will be small.

At this stage, I encourage you to deliberately go too far. Make several small rake adjustments, moving through the level position and slightly beyond it. As you do this, you should notice that the exact level position produces a noticeably lower noise floor and a quieter overall presentation. Return to this position and note the reading on the DigiPas level. Confirm that the azimuth still matches the left speaker.

At this point, you have aligned one degree of freedom—the rake angle—to a grid line. The question is whether this is the *correct* grid line. Once again, the floor complicates matters. If you later determine that you need to move to a different rake grid, the rake change will be relatively large. The angle you measured earlier indicates approximately how far you will need to move the speaker. Moving to a new rake zone will require a new set of compensating moves, beginning with front-to-back alignment.

Finer Front-to-Back: Beethoven Timing

For this round of refinement, we will use left-to-right timing to align the front-to-back position of the right speaker. Play the opening minute of *Beethoven's First Piano Concerto*. During this passage, the orchestra repeats a series of staccato quarter notes. Instruments are distributed across the soundstage, with some on the left and others on the right.

Listen carefully and focus on the timing of the notes from left and right. Ideally, all notes should arrive at your ears simultaneously. In practice, the speaker is unlikely to be exactly on a grid line at this stage. Ask yourself whether you hear *Ba... Ba, Ba, Ba* or *Ba-bup... Ba-bup, Ba-bup, Ba-bup*. If you hear the latter, a front-to-back adjustment is required.

If the right speaker sounds late, bump it forward. If it sounds early, bump it backward. Continue making small adjustments until the notes from the left and right speakers arrive at the same time.

Opening Bars of *Beethoven's First Piano Concerto*

Finer Lateral Adjustment: Lock Leonard's Voice

When listening to **"Treaty,"** notice that Leonard's voice contains two distinct components: the familiar low, gravelly baritone, and a higher-frequency, breathy quality that emerges when he is nearly whispering. For the purposes of lateral adjustment, begin by focusing on the low-frequency portion of his voice. Once that is centered, attention will shift to the higher-frequency component. Be very precise. Leonard's voice should form a circular image, and the center of that circle must lie exactly on the centerline between the two speakers.

At this stage, it is common for the low baritone component to be centered while the higher, breathy component splashes to the left (or right). This indicates that the frequency bands are not yet aligned spatially. To bring the higher frequencies into alignment with the lower ones, we will use **toe-in**.

But for now, our goal is for Leonard's voice to be solidly anchored at the center. If you have difficulty visualizing the exact center, place a piece of painter's tape on the wall and mark the center point. Listen carefully and determine whether his head is precisely centered. Once the low-frequency image is centered, do not assume the image is solid. To test this, move your head—without moving your torso—slightly left and right. Leonard's voice should remain fixed at the center point on the wall. If his voice shifts left when you move your head left, the image is not yet solid. Continue bumping the right speaker until you can move your head left and right without Leonard's voice moving. (Recall the observations on localization uncertainty from De Sena back in chapter 1 where he showed the stereo "sweet spot" is three head widths.)

Finer Toe-In: Match Left/Right Focus

Toe-in is primarily about **focus**. The right side of the soundstage should be just as focused as the left. This can be evaluated using almost any track, but we will begin with **"Till Tomorrow."** Listen to the trumpet blasts on the right, in the center, and on the left. Pay attention to the echo chamber and the clicks. Are they more focused on one side compared to the other? Vocals can also be used for assessing focus. For example, listen again to Kat singing "Lucky." Do you hear a difference between the left and right side of her "head?"

If comparing left and right focus is difficult, try this technique: while the music is playing, lean slightly to the left, then slightly to the right. This makes differences in focus easier to hear.

Now listen to the first 30 seconds of the reference tracks from **Leonard Cohen, Ruth Moody, and Cecile McLorin Salvant.** All three voices should be exactly centered. It is common at this point for Leonard's voice to be centered while Cecile's voice remains slightly pulled to one side.

Listen carefully to Ruth's voice and determine if it is exactly centered. The mid and higher register vocals are giving you toe-in information. If Cecile is pulled left, then the right speaker needs more toe-in. If Cecile is pulled right, then the right speaker needs some toe-out.

If it is not centered, go to the right speaker, hold the front inside spike steady, and use your other hand to bump the lower rear corner of the cabinet to accomplish the necessary toe-in or toe-out. Listen again—Ruth's voice should move toward the center. Repeat this process until her voice is centered.

After each adjustment, it is a good idea to listen again to all three singers. This allows you to hear how each frequency band is responding, particularly the breathy component of Leonard's voice. Once Cecile's voice is centered, the breathy portion of Leonard's voice should also fall largely into place.

Last Compensating Move for this Stage

By bumping the inside rear of the speaker cabinet, we have done more than just change toe-in. First, rotating the speaker moves the acoustic center forward or backward. Second, if the front inside spike remained stationary, the acoustic center also shifted laterally. Listen carefully and make the necessary compensating adjustments to correct for these changes.

Round 3: Final Spatial Cleanup

At this point, we continue refining clarity. Continue using **"Yellow Brick Road"** to confirm that the soundstage remains level. Smaller rake adjustments can also be guided by observations from other recordings. If at any point the soundstage appears higher on the right, rake the right speaker forward.

Even Finer Adjustments for Clarity

At this point, the Shostakovich orchestra should be relatively evenly distributed between the speakers, and the vocalists should be centered, regardless of the frequency range in which they are singing. Granted, this does not necessarily mean that the system is aligned to the nearest lateral grid line.

Figure 39: Congested Noise in the Center

Lateral Movement: Clarity, Perspective, and Piano Size

Image size is a matter of perspective. If you attend a symphony and sit in the first few rows, you will be looking up at the instruments, and they will appear quite large because of your proximity. If you sit closer to the middle of the hall, you are more likely to look straight ahead at the orchestra, and the instruments will appear smaller. Sit farther back still, and their apparent size diminishes even more. As discussed earlier, the rake angle controls vertical perspective: by adjusting rake, you can make images appear higher or lower, one grid at a time. Image size, however, is primarily adjusted through **lateral movement**, which changes the amount of grid overlap between the left and right speakers.

It is now time to set the lateral position of the right speaker. For this, we will again use *Beethoven's First Piano Concerto*. Begin listening at **3:13** and focus on the size and shape of the piano. When the descending arpeggios begin at **3:40**, pay close attention to the direction of the notes. Are the high notes on the left or the right? The high notes should be on the right, and the low notes should be on the left. Additionally, when the image is sized for a realistic, mid-hall perspective appropriate to your room, the piano should sit slightly off-center to the left and should not stretch continuously from one speaker to the other. If you listen carefully and "visualize" with your ears, you should perceive the piano as being positioned diagonally, just left of center. This corresponds exactly to how it is depicted on the album cover and as visible in the illustrations that follow.

As we learned while setting up the left speaker, lateral bumps add clarity to the piano's sound. Listen carefully to the passage from **3:13 to 3:53** and make small lateral bumps to increase space between the notes and to move forward through any excess brightness. As you move the speaker laterally, notice how the apparent size of the piano changes. Initially, the piano may sound enormous, with oversized keys filling the entire space between the speakers. As you continue, it will become smaller, and the keyboard will begin to descend on the left and flatten out as it extends toward the right speaker. With further adjustment, the image will shrink until it sounds as though a piano is sitting between the speakers. You may need to bump the right speaker once or twice simply to switch hands so that the high frequencies remain on the right.

This is a critical moment in the process. How do you know whether you are in the correct rake grid? When Arturo Benedetti Michelangeli plays the descending arpeggios between **3:40 and 3:50**, the keyboard can sometimes sound disconnected. As the notes descend, they may jump in space and sound tonally disjointed, as though there are two separate pianos—one handling the high notes and another handling the low notes. You can attempt to resolve this with small rake and azimuth adjustments, aiming to move the right speaker so that the keyboard coheres into a single, continuous instrument. Granted, if repeated attempts fail to reconnect the keyboard, this is a strong indication that you are in the wrong rake grid. In that case, return to the rake adjustment, move to the next grid, and work through the process again. This is not easy work—but the result, when it finally locks in, is unquestionably worth the effort.

Huge Piano Keys

Piano Begins to Take Shape

Piano is The Correct Size But Wrong Direction

Piano is Correct

Toe-in Refinement

Since we have moved the speaker laterally, we will need to readjust and refine the toe-in. Listen to the trumpet starting from the 1:00 mark once again in **"Till Tomorrow."** This time, focus on the breathy component of the sound. Ask whether that breathiness is centered or streaming to the left or right. If the trumpet's breath is streaming to the right, toe-out slightly more. If it is streaming to the left, toe-in slightly more. The entire trumpet sound should be centered.

Helper Tracks When Imaging Won't Cooperate

If you find yourself struggling to place objects accurately in space, you can turn to these two songs:

1. **"Makes Us Stronger"** by **Ghost Rider**
2. **"Bass Head Jazz" by CeeLo Green**

I discussed the **"Makes Us Stronger"** track in the reference songs section. Because it is electronic music, most of the images it produces are symmetrical. Pay particular attention to the bass pulses, the circular upper-midrange notes at the beginning, and the shakers. Your task here is to use the right speaker to gently push or pull each image on the right side until it precisely matches its corresponding image on the left. The shakers in particular will normally require some azimuth adjustment to correct symmetry.

"Bass Head Jazz" by CeeLo is another electronic track that can be very helpful in diagnosing setup problems. Consult the description of the song in the reference tracks section for a detailed description of what you should hear. Again, if the left and right sides are not mirror imaged you will need to adjust primarily rake and lateral movements to gain symmetry.

Confirm the Fully Developed Sound Field

Earlier, I asked you to feel how the sound presses upon your body. At this stage, we must confirm that the sound field is even and fully developed.

Begin with the bass. Listen to the pulse in **"Till Tomorrow."** It should strike you squarely, with even, centered pressure. Next, listen to the opening of **Shostakovich's Symphony No. 1**, before the music itself begins and the hall slowly comes to life. The ambience of the hall should sound uniform and continuous, filling the entire space between the speakers and extending upward toward the ceiling. Refer to the images in the reference track section that show the "sparkly mist." There should be no gaps or holes in that mist.

There is one final way to assess pressure distribution. Listen to *L'Italiana in Algeri,* this time selecting a particularly loud passage. From approximately **2:20 to 3:00**, the full orchestra is playing. You should feel even pressure across your face. If it helps, insert earplugs so you are not distracted by the sound itself and can focus entirely on the physical sensation.

Stage 3: Temporal Alignment — Time to Get More Precise

At this point, you should have a level soundstage with a strong central image, independent of frequency. Bass impacts—whether from a kick drum or electronic source—should feel centered and even. There should be minimal clashing, and the left and right speakers should sound equally focused. Music will likely be very pleasant to listen to, with nothing poking or irritating the ear. For most people, this would be the natural stopping point.

But we are not finished.

Up to now, the movements we have made were relatively large, intended to bring the speakers very close to a mirrored alignment. We now need to listen more deeply and make extremely fine adjustments to clarify musical timing. These movements will be *very* small—spike rotations of less than **1/64 of a turn**, often so slight that you may not see the threads move at all.

Some coarse right-to-left time alignments have already occurred through the use of Beethoven's First Piano Concerto, but now, as with the left speaker, we must also align the right speaker's drivers. Additionally, we need finer alignment between the right speaker and its mate, the left speaker. This is the most difficult stage of the process—and also the most rewarding. We will use several pieces of music to examine timing.

For front-to-back we will use:

- **"Poem of Chinese Drum"**
- **Rossini's L'Italiano in Algeri**
- **Debussy's String Quartet in G minor**

For driver-to-driver alignment we will use:

- **Beethoven's First Piano Concert**
- **Rossini's L'Italiano in Algeri**
- **Shostakovich's Symphony No. 1 (Allegretto)**

As we move through the different pieces of music the tempo will increase demanding that we have tighter and tighter alignment. For example, in the opening of the Beethoven concerto is played at about **2.5 beats per second**, using half and quarter notes, yielding approximately **0.4 seconds per note**. At the **27-second mark** in Shostakovich, all the strings again play together at the same tempo, but now in eighth notes—only **0.2 seconds per note**.

We will start by coarsely aligning the drivers in the right speaker before proceeding to right-to-left alignment. You have already gained driver alignment experience with the left speaker. We will now execute in the same way on the right speaker.

Listen to the first 30 seconds of *Beethoven's First Piano Concerto*. The violins, cellos, and basses all play together in time. Again, you will hear the basses and cellos positioned on the right, the violins on the left. Listen first to the timing between bass and cellos, then between cellos and violins. Is the bass ahead of the cello, or behind? Are the cellos ahead of the violins, or behind? You will likely need to repeat this passage several times.

The cellos and basses must be aligned so that their plucks arrive at your ears simultaneously. Notably, even the smallest left-to-right misalignment will introduce bass overhang, which blurs the transient and makes the cellos harder to hear. We will need to go back and forth between aligning the right speaker drivers and aligning left-to-right.

We used this piece of music in the previous section. So, the cellos and violins should sound aligned. Focus your attention on the bass and cellos on the right. (Focusing on only one speaker while ignoring the other is a listening skill that must be practiced to learn) If the cellos are late, rake the speaker forward slightly (assuming the tweeter is on top—see the section on driver arrangement). If the cellos are early, then rake the speaker backward. This adjustment should be *extremely* small. Reassess and repeat until the bass and cellos play together. Remember: these spike rotations must be *very, very small.*

Now that the right speaker's drivers are more aligned, we will turn our attention back to the right-to-left alignment. We would need to do this regardless since raking the speaker in either direction has affected the front-to-back position of the drivers.

Return to "Poem of Chinese Drum." The bass drum should be perfectly centered, with a deep, powerful "boom." The large Kodo drums produce significant low-frequency energy, so we expect some bass overhang before the next impact. Listen again to the passage between **0:30 and 1:10**, where the left and right drums are struck simultaneously. Pay special attention to the strike at **0:43**, where alignment is extremely tight. At this stage, the transient attack should be sharply concussive. If you hear the right speaker playing early (Ba-BOOM), then give it a very slight bump backward. If you hear the right speaker arriving late (Ba-BOOM), then give it a very slight bump forward. Repeat this until you hear one large "BOOM" at the 0:43 mark.

Note: If at this point the drum is not perfectly centered then you should go back to the previous section and ensure everything is spatially aligned.

Return to Beethoven's First Piano Concerto and see if you can hear the cello alignment more clearly. Repeat the process of aligning the bass' to the cellos on the right using rake. Then align the cellos to the violins. Once you are pleased with the result we will turn our attention to a much more precise tool.

Listen to the opening 23 seconds of *L'Italiana in Algeri*. This is an outstanding piece for time alignment. This piece is played at approximately **52 bpm**, which corresponds to a string pluck roughly every **0.9 seconds**. The recording is quiet, and the strings are delicately plucked, leaving ample space for each note to decay fully before the next begins. This allows us to hear the precise instant of note attack.

In this passage, the strings play in unison, with violins and violas on the left and cellos and basses on the right. The pizzicato strings should sound lively and bouncy. When properly aligned, the music takes on a playful, almost mischievous character.

Because the strings are playing in unison, you should hear *bap, bap, bap*—not *ba-bap, ba-bap, ba-bap*. If you hear the latter, the left and right speakers are not aligned to the nearest grid line. Determine whether the right speaker is early or late by noting which side you hear first. If the right speaker (cellos and basses) is early, bump it slightly backward. If it is late, bump it slightly forward. Continue adjusting until the pizzicato sounds consistently aligned.

Note: Even after you listen multiple times, you may find it difficult to determine which side is early/late. That is okay. Just go over to the right speaker and give it a very slight bump backward. This changed the timing relationship — either for better or worse. If it is worse, then you now know you went the wrong direction. Remember, as you are first learning this process a great deal will be experimental. You will need to make a change and assess better or worse.

The pizzicato figure repeats four times. If the timing sounds off in any of the first three repetitions, several bumps are likely required. The fourth repetition is the most revealing and easiest to judge. If the first three cycles sound aligned and the last does not, only one or two additional small bumps are usually needed. When you return to the first three cycles, you should notice that they now sound cleaner and more precise than before.

Opening Bars of *Rossini's L'Italiana in Algeri*

Once the strings are aligned, note that any bump will affect both toe-in and rake. Confirm that image height remains consistent from left to right, and that focus is balanced.

At this point, the sound should already be rather wonderful. If you momentarily silence your analytical mind and simply listen to Beethoven, the orchestra may sound exceptional. Still—we are not finished.

Now make one final check of left-to-right timing. Listen to the first four notes of the second movement of **Debussy's String Quartet.** All the strings are plucked together, precisely timed and evenly balanced. You may need to hear these four notes several times. Listen and correct until you are convinced that both speaker's sound arrives at your ears at exactly the same moment.

If the right speaker is early or late, make a correction. You may either:

- Tap the speaker forward or backward, **or**
- Adjust the rake slightly forward or backward

Either method will require only the tiniest movement.

Some guidance:

- If the sound is slightly bright and the right side is late, tap the right speaker forward.
- If the right side is high and late, rake it forward.
- If the right speaker is early or late and you hear slight bass overhang in the opening of *L'Italiana in Algeri,* bumping is usually preferable to raking.

There are many possible combinations, but the essential point is this: you are now working at a level where each adjustment affects multiple attributes. When choosing a movement to fix *A*, look for a *B* that can also benefit, so that one adjustment serves two purposes.

Claude Debussy's *String Quartet in G Minor*

Note: If the Beethoven or Shostakovich orchestra sounds "noisy", especially in the center, this indicates a lateral grid misalignment. Give the right speaker a very slight lateral bump and reassess. If the instruments sound more clearly defined, then you are going the right direction. Give another very slight lateral bump and assess once again. Keep doing this until you are happy with the result.

***Shostakovich's Symphony No. 1* (at 31 seconds)**

Finer alignment can be achieved with Shostakovich, at approximately the **3:48 mark** a single violin plays. This recording uses two microphones, and the delay between them is audible. The primary microphone captures the violin in the left channel, since the violins are positioned to the left, while the secondary microphone captures the same instrument and feeds the right channel. We can use this characteristic to refine timing even further.

Focus on the decay of the violin tone. The easiest place to hear this is when the violin sustains a note. Our goal is to match the decay on the left and right sides. To do this, use a rake angle adjustment, as any bump would be far too coarse at this stage. The decay on the right side will most likely be slightly too long. Go to the back of the right speaker and barely break the thread on the rear outside spike, raising it by an imperceptible amount. Listen again and repeat as necessary until the decay of the violin matches on both sides.

Finishing Adjustments: Finally Azimuth

We will now finish with small azimuth adjustments. Queue up *Schubert's Piano Sonata No. 20, Andantino*. This is a very slow, deliberate piano piece that gradually ascends the scale. As written in the score, Mr. Andsnes plays many dyads and chords. These should sound extremely pure in tone. Often, however, the notes will exhibit a subtle dissonance, sounding slightly billowy or ringy. If you hear this, move to the right speaker and make the smallest possible azimuth adjustment by raising the inside spikes. Listen again. You should hear the lower notes become purer, while the higher notes may still retain some ring or dissonance. Make another extremely small azimuth adjustment—again barely breaking the thread. Continue this process as needed until even the highest notes sound clean and pure. Reference the description in the reference songs for more detail on other useful moments during Shubert's Sonata.

Except from *Schubert's Piano Sonata No. 20 Andantino*

This sonata is played at a very slow tempo, with great care taken to allow each note to decay fully. It should never sound hurried or rushed. As the noise floor drops, the music will seem to slow down. One effective way to achieve this final sense of ease is with the slightest additional toe-in adjustment. This often works because even a tiny increase in toe-in can significantly lower the acoustic noise floor of the room, far more than the magnitude of the movement would suggest.

Here is a brief anecdote. You may recall the two world-class string players mentioned earlier in the book. At the end of a setup session, we were listening to Mr. Andsnes play *Schubert's Piano Sonata No. 20*. Following the approach described above, the speaker was given the tiniest amount of toe-in. The reaction from both musicians was immediate and intense. They were stunned, exclaiming, "W—T—F just happened?! How did you just change the tempo of the music?" This brief story illustrates the power of subtle adjustments to lower the noise floor.

Once you are satisfied with Schubert, return to Shostakovich and check the "room pop" at approximately **2:33**. If it is not fully developed, give the right speaker the slightest outward bump. In most cases, this final adjustment will lock everything into place.

Queue up *Shostakovich's Symphony No. 1* one final time. You may listen up to **2:25** and simply enjoy the music, or you can fast-forward directly to that point. Beginning at approximately **2:33**, you will hear a back-and-forth exchange between the plucked bass and the horns (Section 1, marked in the score below), followed by a similar exchange between the bass and the violin (Section 2, also marked below).

The bass is playing an **A2 flat**, which is approximately **104 Hz**. This note should be very quiet, with no bass overhang. There should be a striking silence between the bass and the violin entries. Any excess bass overhang will destroy this effect. When it is correct, it can feel as though all the air is being pulled out of the room, followed by a subtle "pop" of the room with each note.

If this effect is not fully realized, it is often the smallest outward bump of the right speaker that brings everything into alignment. Do not be discouraged if you are unable to achieve this on your first pass through the process. Reaching this point represents a genuine level of mastery.

Congratulations—you've arrived. Play several songs you know well. The bass should be smooth, articulate, dynamic, and punchy. The highs should be fully present without sounding fatiguing, bright, or edgy. Stereo imaging should be rock solid, with each instrument occupying a clearly defined place in the space between the speakers. When the recording allows, the soundstage should extend from wall to wall and from floor to ceiling. The musicians should sound tonally pure and perfectly in time with one another. What you hear now should feel like a real musical performance unfolding in your room.

Recordings in which the artist has deliberately played with phasing should be especially entertaining. For example, in CeeLo's "Bass Head Jazz," there is a passage at about **1:04** that can sound as though it is circling around your head.

One thing you should notice is how good the sound is anywhere in the room. It sounds calm and relaxing to listen to music. If you have visitors to listen to your system, you will undoubtedly notice how sitting to one side of the main listening position still yields amazing imaging. You can still hear the singer in the center. The orchestra will still be spread out between the speakers. All of this will allow anyone in the room to enjoy the music.

At this point, you should be able to picture yourself—listening to your favorite music, completely relaxed.

Summary: The System Becomes One

In Chapter 9, we completed the process of bringing the left and right speakers into alignment so that they function together as a single coherent acoustic source. Building on the reference established with the left speaker, we carefully introduced the right speaker and adjusted its position until both speakers worked together in precise timing and spatial harmony.

This step transformed the system from two independent sound sources into a unified stereo instrument. When the alignment is correct, the speakers themselves begin to disappear, replaced by a stable soundstage in which voices and instruments occupy believable positions in space. Imaging becomes centered and stable, tonal balance improves, and the system begins to reveal the spatial relationships embedded in the recording.

Equally important, this chapter demonstrated that stereo alignment is not achieved through simple symmetry. Instead, it emerges through careful listening and deliberate adjustments that respect the timing relationships between the speakers, the listener, and the room. Small positional changes, often measured in millimeters, can dramatically influence coherence, clarity, and dynamic expression.

At this point in the setup process, your system should be capable of producing a stable phantom center, clear separation between instruments, and a soundstage that extends naturally across and beyond the speakers. The room, the speakers, and the listening position are now working together as a unified system.

However, this is not the end of the journey. The skills you have developed—careful listening, patience, and precise adjustment—will continue to deepen as your experience grows. As you spend more time working with your system, you will begin to discover additional musical cues that guide small refinements in speaker placement and system balance.

With the main speakers now aligned and stereo coherence established, the foundation of the Precision Setup Method is complete. The next step is integrating subwoofers to extend this coherence into the lowest frequencies. The following chapter begins that process.

CHAPTER 10

SUBWOOFERS AND BASS TOWERS — EXTENDING THE FOUNDATION

You may wish to augment your speakers' bass with subwoofers, or your speakers may already include separate bass towers. Either way, these low-frequency devices must be integrated with the main speakers using the same principles we applied to the mains. **Do not begin subwoofer or bass-tower integration until the main speakers are fully positioned and optimized.** Adding another woofer will not "fix" an imperfect setup—it will amplify existing problems. If you are not **100% satisfied** with how the main speakers sound on their own, return to the main setup and refine it until the system is performing as well as it can before introducing additional bass sources.

A subwoofer is about far more than extending low-frequency response. Yes, it can deliver that super-low bass that is undeniably fun—especially in EDM. But when properly integrated, a subwoofer also increases the sense of **space** and can **widen the perceived sound field**, which is often more musically meaningful than sheer low-end weight. The lowest octave also adds **body** to instruments and increases the listener's sense of **envelopment**. Hearing a solo oboe or flute with and without a properly integrated subwoofer can be startling. How can energy below 30 Hz influence what we perceive from a flute playing around 1000 Hz? Psychoacoustics? Perhaps. I don't pretend to know the full mechanism—but the effect is real, and it is one of the great rewards of doing this correctly.

This chapter is substantial, with a fair amount of information to absorb and several technical concepts to understand. That is to be expected, because in the process of integrating subwoofers you are, in effect, taking on part of the loudspeaker designer's job. You will choose the subwoofer driver and enclosure approach, select a crossover point, and then align **gain (volume level)** and **phase** so the subwoofer behaves as a coherent extension of the main speakers. If that sounds daunting, don't worry—each step you learn brings you closer to a seamless result. **Read this chapter twice before beginning.** The first pass provides a full overview. The second pass will be clearer because you'll already understand where the process is going.

In the tradition of giving you a broad map before we begin, here is the outline of this chapter:

- **Preliminaries: choosing subwoofers**
- **Setting gain and crossover**
- **Placement**
- **Phase and group delay**
- **Finding the right phase setting**
- **Integrating a subwoofer**
- **Hypothetical case study**

Which Subwoofer

Let us assume you wish to add a separate subwoofer to your system. A brief search will reveal an overwhelming number of options on the market, the majority of which are designed primarily for home-theater use. There is nothing inherently wrong with that, but the requirements of home theater and music reproduction are profoundly different. Home theater subwoofers are expected to reproduce spectacle—the subterranean rumble of a subway train or the concussion of depth charges exploding beneath a submarine. Music, by contrast, demands **nuance**. In the bass, we are listening for **texture, articulation, and timing**. Yes—even below 40 Hz, these qualities are plainly audible in a high-resolution audio system. Achieving them requires a subwoofer whose speed and articulation can keep pace with the main speakers.

The first major consideration is **driver size**, and here we encounter two distinct use cases. In the first, the subwoofer is intended to extend the bass of a small two-way or compact three-way loudspeaker. In this scenario, a subwoofer using an 8-inch or 10-inch driver can be entirely appropriate. In the second case, the system already employs large, full-range loudspeakers that extend cleanly below 40 Hz. Here, a 12-inch driver should be considered the absolute minimum.

The reason is straightforward: the lowest frequencies require significant air movement to be reproduced convincingly. As driver diameter decreases, the cone must move farther to achieve the same sound pressure level. Increased excursion leads directly to increased distortion, and the farther the cone travels, the longer it takes to stop moving. The result is bass overhang and temporal blur—precisely what we are trying to avoid. For music reproduction, we do not want large excursions; we want **efficient air movement**. For this reason, 15-inch or even 18-inch drivers are the better choice. An 18-inch driver moves approximately **2.25 times as much air** as a 12-inch driver for the same amount of cone excursion.

That said, driver size alone does not guarantee success. A poorly designed 18-inch woofer can produce more distortion and overhang than a well-engineered 12-inch driver. The qualities we seek in a subwoofer driver are no different from those we value in the main speakers. The moving mass should be as **light and stiff** as possible. The surround should allow free motion without resisting or storing energy. The driver should remain free from breakup anywhere near the operating range in which it will be used.

Unfortunately, many subwoofers employ heavy cones and thick, stiff surrounds. To my eye, this is a clear indication of a design aimed at home-theater duty. In such cases, the manufacturer relies on a high-power, low-cost plate amplifier to force the driver into motion. That approach holds no appeal for musical reproduction. I would far rather see a **100–200-watt amplifier** paired with an **easy-to-drive, responsive woofer** than a brute-force solution built around sheer wattage.

The next consideration, after the driver itself, is the **cabinet design**. Subwoofers are available in a wide variety of enclosure types: sealed, ported, passive radiator, open baffle, panel, horn-loaded, and even rotary designs. While each has its place, most listeners seeking musical bass are typically looking for a conventional *cone-in-a-box* solution. For that purpose, the **sealed enclosure** offers the highest likelihood of success.

Assuming competent engineering, a sealed enclosure will exhibit **significantly lower group delay** than ported or passive-radiator designs. (Group delay will be discussed in detail later in the section on subwoofer phase.) This is not to suggest that a ported subwoofer cannot be designed to integrate well with a high-resolution speaker—it certainly can. However, unless your loudspeaker manufacturer offers a purpose-matched subwoofer, a sealed design should be your first consideration when searching independently.

Regardless of enclosure type, **cabinet rigidity is essential.** The internal pressures generated by low-frequency reproduction are substantial, and insufficiently braced cabinet walls will flex and resonate. If allowed, this stored energy will smear transients and muddy bass articulation. Proper internal bracing and resonance control are not optional; they are fundamental to clean, intelligible low-frequency reproduction.

The final major decision concerns whether the subwoofer should be **active or passive**. The vast majority of subwoofers on the market are active, meaning the amplifier and crossover are built directly into the enclosure. This approach is undeniably convenient: a compact, self-contained solution requiring only a single interconnect cable. Convenience, however, comes at a cost.

The principal drawback lies in the amplifier itself. Most active subwoofers rely on inexpensive class-D plate amplifiers built to meet aggressive price targets. These amplifiers typically generate significant electrical noise. Because the plate amplifier is connected directly to the preamplifier, much of this noise is fed back into the main system, contaminating the midrange and treble. Earlier in this book, we discussed the destructive effects of noise in detail. By introducing an active subwoofer, you have effectively added a **noise generator** to the system—something that demands careful consideration.

For these reasons, **passive subwoofers** deserve serious consideration in high-resolution audio systems. While this approach requires the purchase of an external active crossover and a dedicated amplifier, it avoids the noise issues inherent in most plate amplifiers. More importantly, it gives you control over the amplification and cabling driving the subwoofer, allowing for far better tonal integration.

The goal is to match the character of the subwoofer's bass to that of the main speakers. As a general rule, the subwoofer amplifier should share the same **topology** as the main amplifier. If your main speakers are driven by a class A/B amplifier, the subwoofer amplifier should also be class A/B. If your main system uses tubes, a high-quality class-A solid-state amplifier is often an excellent complement. As with every other aspect of the system, thoughtful component matching in the subwoofer chain plays a critical role in achieving the sense of coherence and musical authority—the *WOW effect*—that we are ultimately pursuing.

How Many Subwoofers

Before we discuss *how* to integrate a subwoofer, we must first address **how many subwoofers are required**. Opinions on this subject vary widely, and there are several commonly proposed methods for "fixing" bass. In my view, however, there is a single correct answer for two-channel music reproduction: **two subwoofers**. Let us briefly examine the merits and shortcomings of using one, three, four, or more.

The sole advantage of a single subwoofer is cost—you only need to purchase one. If you choose to use a single subwoofer, there is only **one viable placement**: directly centered between the main speakers. This is the only location where a single subwoofer can pressurize both sides of the room evenly. If the subwoofer is placed in a front corner, for example, the resulting imbalance in sound pressure becomes immediately obvious and will skew the sound field. This also answers a common question: *Should I buy one expensive subwoofer or two less expensive ones?* The answer is clear—**buy a stereo pair at the best price point you can afford**.

Using more than two subwoofers serves a different purpose altogether. The goal in that case is not stereo integration but rather smoothing the bass response over a large listening area—a technique known as a **distributed bass array**. This approach places three or four subwoofers in widely separated locations throughout the room so that the modal nulls produced by one subwoofer are filled in by another. When implemented correctly, this method works exactly as intended and can produce an impressively smooth low-frequency response.

That said, if bass smoothing is the primary objective, much of this benefit can be achieved with **two subwoofers combined with DSP**. The question of how many subwoofers are actually required for a distributed array was addressed rigorously in a well-known paper by **Todd Welti**. His research demonstrated that adding subwoofers beyond four does not meaningfully improve frequency-response smoothness. In practice, three subwoofers yielded results nearly indistinguishable from four. This approach is well suited to **home-theater systems**, where smooth bass across a wide seating area is desirable. It is largely unnecessary for **two-channel stereo**, where there is only a single critical listening position.

I appreciate that this may sound theoretical, and that some readers will want evidence. Below is an example from a real installation: a pair of two-way speakers augmented by a stereo pair of 12-inch subwoofers positioned just behind and slightly outside each main speaker, in a relatively small room. The resulting in-room measurement in Figure 40 shows a remarkably flat response from 20 Hz up to the 80 Hz crossover point. This is not simulated data—it is a real, measured result.

Figure 40: Frequency Response of 2-Way Speakers with a Pair of Subwoofers

(1/24 Octave Smoothing)

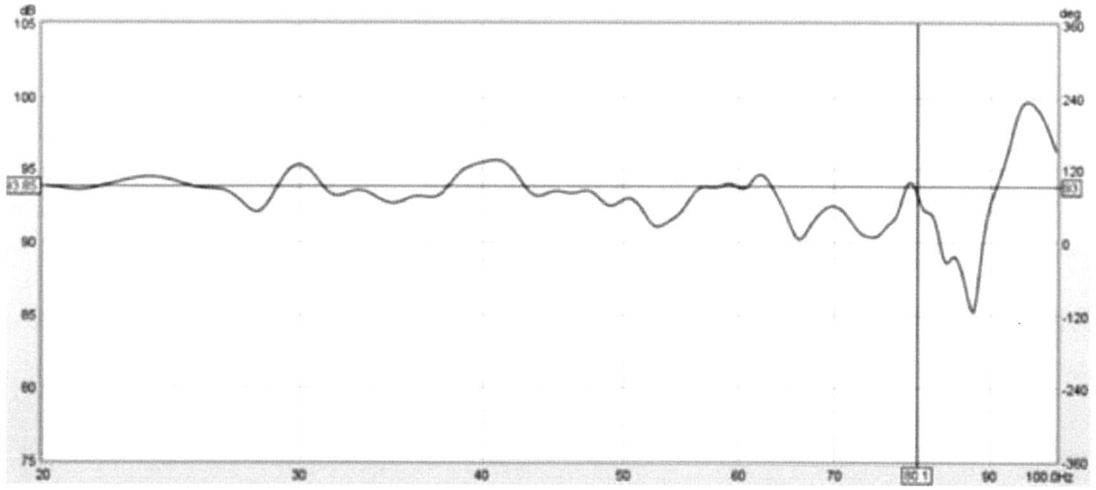

One subwoofer manufacturer advocates stacking subwoofers—up to three units—beside each main speaker. This approach creates a stereo pair of subwoofer stacks and allows independent control of crossover frequency and gain for each of the six individual drivers. One advantage of using multiple drivers is that each driver operates at a much lower gain than it would in a single-subwoofer-per-side configuration. Lower gain reduces cone excursion, which in turn decreases settling time and distortion.

If you choose to pursue this approach, the stacked subwoofers must be integrated exactly as you would any other bass tower, following the same principles outlined in this chapter. That said, before committing to a system with six subwoofers, I strongly recommend starting with two. Integrate them properly, live with the result, and only then decide whether additional subwoofers are truly necessary.

Preliminaries

Whether you choose an active or passive subwoofer, some cabling will be required. Most subwoofers on the market today include built-in amplification and control electronics and accept **low-level inputs** from the preamplifier via RCA or XLR connections. You will therefore need a **pair of interconnects** long enough to place the subwoofer in each of the candidate locations you plan to evaluate and to connect it to the active output of your preamp.

Before purchasing cables, measure the distance to **every potential placement location**. You will likely find that different positions require significantly different cable lengths. At this stage, the interconnects do not need to match those used elsewhere in your system. Simply purchase a **long, inexpensive pair** that can comfortably reach all candidate locations. Once you have determined the final subwoofer position, you can then focus on selecting a higher-quality interconnect that complements the rest of your system. Ideally, this would be the same model used between your other components, but if that option is too expensive for subwoofer use, stay within the same manufacturer's line and choose a more affordable alternative.

Before positioning the subwoofer near a wall, take time to **examine the rear panel** while it is still out in the open. It is much easier to familiarize yourself with the connections and controls when they are fully visible. Note the orientation of the power-cord inlet and the location and direction of the RCA or XLR inputs. Taking a quick photo can be helpful as a reference, since access to the rear panel will be limited once the subwoofer is close to a wall.

Control layouts vary, but most subwoofers provide **three rotary controls and a switch**: gain, crossover, phase, and a polarity switch. (See following illustration.) Begin by setting the crossover slightly below the frequency at which the main speakers' bass output begins to roll off—technically, the point where their output is down by 3 dB. Set the gain so that the subwoofer's output roughly matches the level of the main speakers just above the crossover frequency. Set the polarity switch to **0 degrees** and place the variable phase control at **90 degrees**, which serves as a neutral midpoint.

These settings are only a starting point. If you want a more precise phase alignment, continue reading before evaluating placement options. Once the subwoofer's final position has been established, the correct gain, crossover, and phase settings will become clear and can be finalized with confidence.

When integrating subwoofers, the addition of **gain, crossover, and phase controls** introduces several new degrees of freedom. With so many adjustable variables, it is hardly surprising that many listeners abandon the process altogether. The goal of the sections that follow is to remove the mystery and provide a clear, repeatable path toward successful subwoofer integration.

Where and How to Place Subwoofers

Now that you have purchased and unboxed your two subwoofers, the next question is where to begin. The first step is deciding **where to place them**. You will likely have several viable options; start by identifying the positions that are practical in your room and work outward from there. Common locations to evaluate include:

1. Outside and beside the main speakers
2. Directly behind the main speakers
3. In the front corners behind the main speakers
4. Along the front wall, approximately **25% of the room width** in from each side wall

When selecting candidate positions, it is important to consider **speaker boundary interference response (SBIR)**. If this concept is unfamiliar, refer to the chapter on room acoustics for a detailed explanation. Our goal is a **smooth frequency response** from the subwoofer with no deep nulls up to the crossover point.

We can estimate SBIR-related nulls in the 20–40 Hz range to define acceptable placement distances. For typical room conditions, the first SBIR null occurs at approximately:

- **14 feet** for 20 Hz
- **9.5 feet** for 30 Hz
- **7 feet** for 40 Hz

This means that, to avoid a null between 20 and 40 Hz, the subwoofer should be placed either **closer than 7 feet to the wall** or **farther than 14 feet from the wall.** This consideration is especially relevant in dedicated listening rooms where speakers are pulled well into the room. In some cases, this information alone may eliminate one or more candidate locations. You are free to experiment regardless, but such positions would not be my first choice.

Each candidate position must be evaluated **individually**. We are looking for the location that produces the **smoothest bass response** from 20 Hz up to the crossover point. However, it is equally important to examine the response **above** the crossover frequency. For example, you may find a position that is very smooth from 20 to 30 Hz with a 30 Hz crossover, but that same location may produce a strong peak at 50 Hz due to a room mode. That peak would need to be addressed through placement changes or other means.

If another candidate position offers similarly smooth response below the crossover point *without* introducing a prominent peak above it, that position should be preferred. In practice, the best placement is usually the one that requires the **least correction** while delivering smooth, well-controlled bass across the relevant frequency range.

Choosing Initial Gain and Crossover Setting

We now need to determine how to choose an appropriate **crossover point** and how to evaluate whether the bass response is smooth from **20 Hz up to that crossover frequency**. Conveniently, the same methods apply to both tasks.

To select the crossover point, we must first identify where the **low-frequency output of the main speakers begins to roll off** and note the loudness at that frequency. There are two practical ways to do this. The most accurate method is to measure the speaker output using software such as REW. This allows you to see exactly how loud each frequency is across the spectrum.

For example, in the frequency-response graph labeled *"Final Spot,"* the output of the main woofers begins to fall off at approximately **35 Hz**. Based on this observation, the subwoofer crossover should be set slightly below that point—around **33 Hz**. The subwoofer gain should then be adjusted so that its output at the crossover frequency measures approximately **90 dB**, which corresponds to the average loudness of the main speakers.

That same graph also shows the final subwoofer response plotted alongside the main speakers' output. We will return to this shortly, as it provides important insight into how placement, crossover, and gain interact.

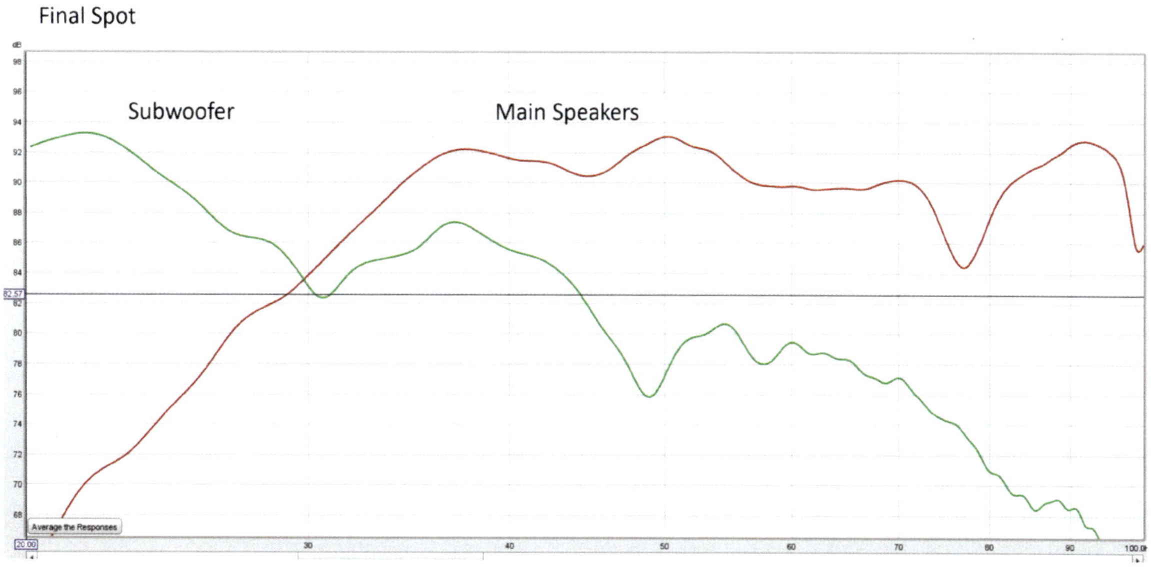

An alternative approach is to use **test tones and a loudness (SPL) meter**. Several vendors offer CDs containing calibrated bass tones, and similar test-tone tracks are available on some streaming services. A simple and accessible option is YouTube, which provides tones at individual frequencies—search for a specific tone such as "*33 Hz test tone* "and play it.

Regardless of the source, the objective is to create a **rough approximation of the frequency-response graph** discussed earlier. Play each test tone in sequence—20 Hz, 25 Hz, 30 Hz, 35 Hz, and so on—and record the SPL reading at the listening position for each frequency. When you examine the resulting series of measurements, it should be clear where the low-frequency output of the main speakers begins to drop off and what the corresponding loudness is at that point.

Once you have identified the roll-off frequency and its level, adjust the subwoofer accordingly. Set the crossover to that frequency (or slightly below it), then adjust the subwoofer's gain so its output is approximately equal to the main speakers' level at the crossover point.

Keep in mind that **phase has not yet been adjusted**. If the subwoofer is partially out of phase with the main speakers, its output will be reduced due to cancellation. After proper phase alignment is achieved, you will almost certainly need to **lower the subwoofer's gain slightly** to maintain the correct level balance.

The Impact of Subwoofer Placement

We need to talk about **subwoofer placement**, because its impact on the final result cannot be overstated. The difference between a poor location and a good one is dramatic. While it can be difficult to convey in words just how bad the wrong position sounds—or how right the correct one sounds—the distinction becomes immediately obvious when viewed through **frequency-response plots**.

The plots below show four sample placements from a subwoofer integration I performed. These plots display **only the subwoofer's response**, isolated from the main speakers. I evaluated three candidate locations: beside the speaker (outside), beside the speaker (inside), and in the front corner. The upper-left plot shows the result with the subwoofer positioned to the **outside of the left main speaker**. A similarly poor result was obtained with the subwoofer placed on the inside of the speaker, so that plot is not shown.

With the subwoofer positioned outside the main speaker, two problems are immediately apparent. First, there is **very little output between 20 and 30 Hz**—exactly the frequency range where we are trying to gain extension and authority. Second, there is a **large peak centered around 55 Hz**. Together, these issues make this placement a textbook example of an extremely poor subwoofer location.

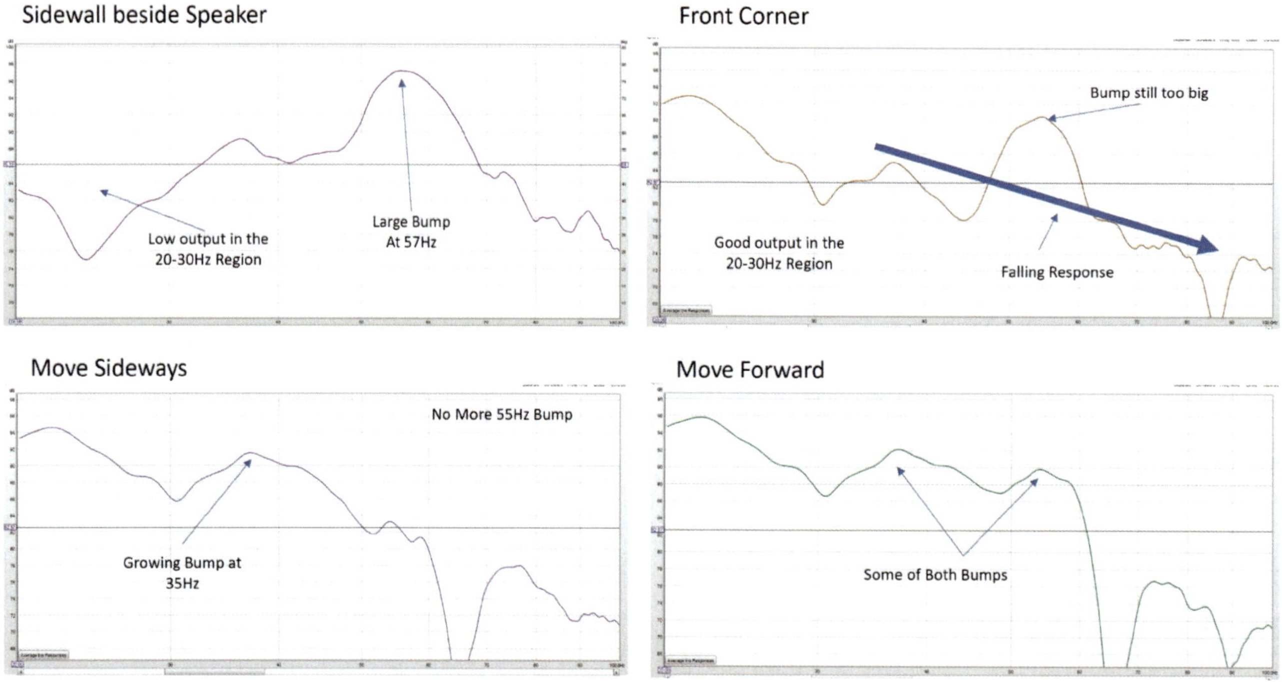

Moving the subwoofer to the **front corner behind the main speaker** produced the low-frequency output we were missing in the 20–30 Hz range. The response now showed a smooth downward slope, which is generally desirable, but a **large peak remained at 55 Hz**. If DSP were available, that peak could be reduced electronically.

Although, even when DSP is an option, it is always preferable to start with a placement that requires **the least possible correction**. Good positioning should do most of the work.

Fortunately, I had the flexibility to pull the subwoofer **out from the front corner by several feet**. To reduce the 55 Hz peak, the subwoofer needed to be moved **laterally toward the center of the room**. As shown in the bottom-left plot, this adjustment successfully lowered the 55 Hz peak—but introduced a new peak at **37 Hz**. At that point, placement became a matter of fine increments. I moved the subwoofer forward by an inch, then sideways by an inch, repeating this process back and forth until the response reached the best overall balance.

The bottom-right plot shows one such position, where both peaks are present but reduced. If forced to choose, I would accept a **small peak at 37 Hz**, provided everything above it continues to fall smoothly. That proved to be the better compromise. Finding this position took **two to three hours**, and the time invested was well worth it. I strongly encourage you to devote the same care and patience to locating the best possible placement for your subwoofer.

The plot labeled **"Final Spot"** shows the subwoofer response achieved using **placement alone**, without any electronic correction. While it is not mathematically perfect, it is highly effective. Several observations stand out. First, there is strong output at **20 Hz**, with a response that gently slopes downward as frequency increases, and no major peaks above **45 Hz**. Second, there is a small peak at **37 Hz**, approximately **5 dB lower than the main speakers** at that frequency. This is not problematic, as it will add only about **1.2 dB** to the combined response of the mains at that point (see the table in the appendix).

Based on the main speakers' -3 dB roll-off, the theoretical crossover point should be around **33 Hz**. In practice, I set the subwoofer crossover to **30 Hz**. Increasing the crossover would have raised the level of the 37 Hz peak, which I wanted to avoid. This illustrates an important principle: crossover settings often involve trade-offs. The controls exist to be used, and the correct settings are those that produce the **best overall musical result**, not the most attractive graph.

Subwoofer Phase

The following sections address **phase**. This topic is admittedly complex—more complex than it should be—but it is essential to understand if subwoofers are to be integrated successfully into a two-channel audio system. The first part explains the fundamental concepts of phase and demonstrates how measurements can be used to identify the correct setting. Don't worry, after that, I will show you how to arrive at the same result using your ears and a few straightforward listening techniques.

Understanding the Units of Phase

Before we can determine the correct setting for the phase control, we need to understand the **units used to describe phase** and what they mean. One reason phase is often confusing is that the phase difference between two sound waves can be expressed in **three different units**: distance, time, or degrees. How can a single physical parameter be described in three ways? The answer is simple: these units are directly related.

A sound wave is defined by three fundamental properties: velocity, frequency, and wavelength. Let's briefly define each one.

- **Velocity** is the speed at which sound travels through a given medium. It depends on how quickly the molecules of that medium can vibrate. In air at in-room conditions, the velocity of sound is essentially constant at about **1,130 feet per second.**
- **Wavelength** is the physical distance a sound wave travels while completing one full cycle.
- **Period** is the amount of time it takes for a wave to complete one full cycle.
- **Frequency** is the inverse of the period (1 / period) and is measured in **Hertz (Hz)**. We usually visualize frequency as how fast the wave oscillates up and down.

These quantities are linked by a simple relationship:

velocity = wavelength × frequency

Because the velocity of sound in air is fixed, knowing the frequency allows us to calculate the wavelength, and vice versa. For example, the wavelength of an **80 Hz** sound wave is found by dividing the velocity by the frequency:

1130 ÷ 80 = **14.1 feet**

So an 80 Hz sound wave is approximately 14.1 feet long.

To build intuition, let's begin by thinking about phase difference in terms of **distance**, which is often the easiest way to visualize it. Imagine a main speaker and a subwoofer, with the subwoofer placed **one foot behind** the main speaker. In this arrangement, we now have two sound sources producing the same frequency.

If we could see the sound waves and measure the distance between their peaks, we would observe that the wave from the subwoofer lags behind the wave from the main speaker by one foot. In other words, the two waves are **out of phase by one foot**. This physical offset is one way to express phase difference.

In the illustration below, wave **B** represents the output from the main speaker, and wave **A** represents the output from the subwoofer. Ideally, these two waves would be perfectly aligned—overlapping so completely that they are indistinguishable. Instead, wave B leads wave A. That spatial offset is the **phase difference** between the two waves.

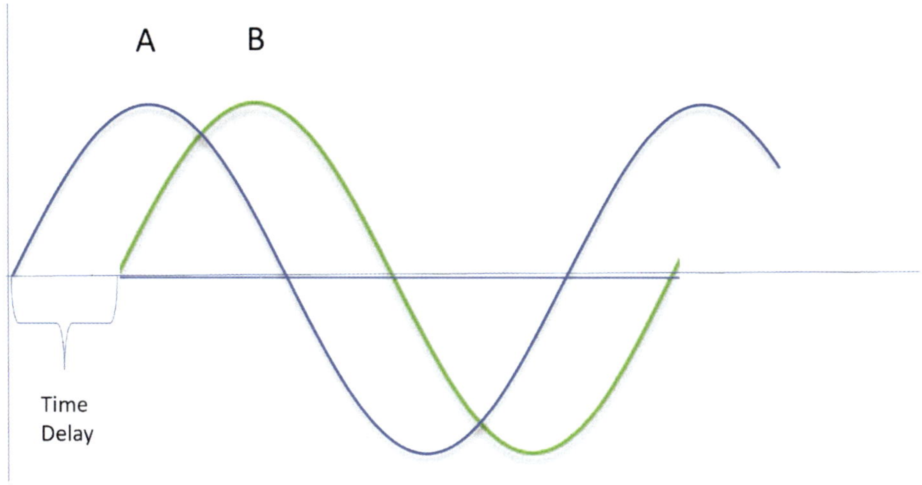

Measuring Phase in Time and Degrees

How do we express phase difference in **units of time**? We can calculate the time it takes for sound to travel a given distance using the basic relationship:

velocity = distance / time

Rearranging the equation gives:

time = distance / velocity

Returning to our earlier example, where the subwoofer is positioned **one foot behind** the main speaker, the distance offset is one foot, and the velocity of sound is approximately **1,130 feet per second**. Dividing the distance by the velocity gives:

1 ÷ 1130 = **0.000885 seconds**, which is **0.885 milliseconds**, or **885 microseconds**.

We have now expressed a **one-foot phase offset as a time delay of 885 microseconds.**

Many modern subwoofers use **digital signal processing (DSP)** rather than rotary knobs. When accessed via a laptop or app, these systems typically provide a field labeled *delay* rather than *phase*. The user is usually given a choice of units—distance (feet or inches) or time (milliseconds). Subwoofers that use rotary controls instead typically express phase in **degrees**.

So how do degrees relate to time?

Recall that the **period** of a sound wave is the time required to complete one full cycle, and that:

Period = 1 / frequency

Using an **80 Hz** example:

Period = 1 ÷ 80 = **0.0125 seconds**, or **12.5 milliseconds**

This means an 80 Hz wave completes one full cycle every 12.5 ms. A complete cycle corresponds to **360 degrees** of phase rotation. Dividing the period by 360 gives the time represented by one degree of phase:

12.5 ms ÷ 360 = **0.0347 milliseconds**, or **34.7 microseconds per degree**.

If the crossover is set to 80 Hz and the phase control is adjusted by **10 degrees**, the corresponding time delay is:

10 × 34.7 μs = **347 microseconds**

If the phase control is set to **90 degrees**, the delay is:

90 × 34.7 μs = **3.1 milliseconds**

In this way, phase expressed in degrees is simply another way of expressing **time delay**.

Here is where phase adjustment becomes unintuitive—and this point is critical. **The meaning of a degree of phase depends on frequency.**

Suppose the crossover is set to **40 Hz** instead of 80 Hz. The period of a 40 Hz wave is:

1 ÷ 40 = **0.025 seconds**, or **25 milliseconds**

Dividing by 360 degrees gives:

25 ms ÷ 360 = **69.4 microseconds per degree**

At 40 Hz, a 90-degree phase shift now corresponds to:

90 × 69.4 μs = **6.25 milliseconds**

Comparing the two examples makes the key point unmistakable:

How far you turn the phase knob (in degrees) to align bass waves depends on the crossover frequency.

This bears repeating:

How far you turn the phase knob (degrees) to align the bass waves depends on the crossover frequency.

To put this into physical terms, imagine a phase control that allows adjustment in precise 1-degree increments. At **80 Hz**, each degree corresponds to **34.7 microseconds**, which translates to approximately **0.039 feet**, or about **half an inch**. At **40 Hz**, each degree corresponds to **69.4 microseconds**, which is roughly **0.078 feet**, or nearly **one inch**.

Even if it were possible to hit each 1-degree increment perfectly, the physical alignment error at 40 Hz would still be nearly an inch. This explains why acoustic measurements often show little visible difference between adjacent phase settings, while the **perceived timing and impact of bass transients can change dramatically.**

The table below summarizes these relationships, showing frequency, period, wavelength, half wavelength, quarter wavelength, and the time and distance represented by **one degree of phase rotation**. It serves as a useful reference when working through phase alignment.

Figure 41: Table for Phase Rotation in Degrees

Freq (Hz)	Period (ms)	WL (ft)	WL/2 (ft)	WL/4 (ft)	1 Degree	1 Degree
30	33.3	37.7	18.8	9.4	0.093 ms	1.26 in
40	25.0	28.3	14.1	7.1	0.069 ms	0.94 in
50	20.0	22.6	11.3	5.7	0.056 ms	0.76 in
60	16.7	18.8	9.4	4.7	0.046 ms	0.62 in
70	14.3	16.1	8.1	4.0	0.040 ms	0.54 in
80	12.5	14.1	7.1	3.5	0.035 ms	0.47 in
90	11.1	12.6	6.3	3.1	0.031 ms	0.42 in
100	10.0	11.3	5.7	2.8	0.028 ms	0.37 in

Visualizing Phase Difference

Let's assume we want to cross a subwoofer to the main speakers at **30 Hz**. At that crossover frequency, the sound wave produced by the subwoofer must be **perfectly in phase** with the wave produced by the mains. If we were to play a pure 30 Hz sine wave, the peaks and troughs from both sources would need to line up exactly for proper summation.

A single, complete sine wave represents **one full wavelength**, which corresponds to **360 degrees of phase rotation**. At 30 Hz, the wavelength can be calculated by dividing the speed of sound (1130 ft/s) by the frequency (30 Hz):

1130 ÷ 30 ≈ **37.6 feet**

This means a 30 Hz wave is approximately **37.6 feet long** from the start of one cycle to the start of the next. Any phase misalignment between the subwoofer and the main speakers at this frequency can therefore be understood as some fraction of that 37.6-foot wavelength—or, equivalently, some fraction of 360 degrees of phase rotation.

Figure 42: 30Hz Sine Wave

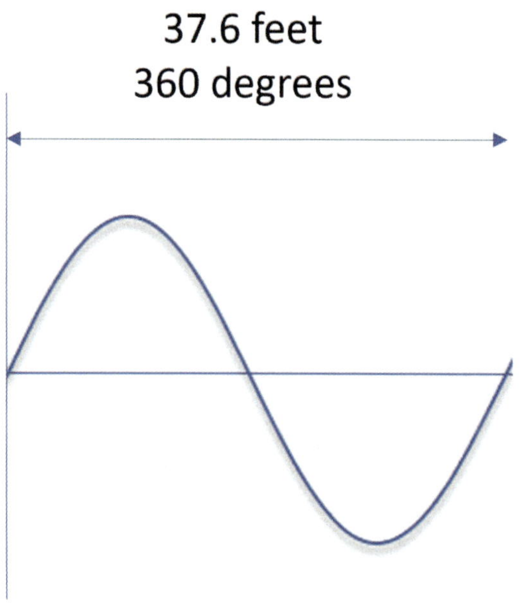

Now visualize two **30 Hz** sound waves: one produced by the main speaker, and one produced by the subwoofer. If the subwoofer is positioned **37.6 feet behind** the main speaker, no phase adjustment is required to align the two waves. By the time the subwoofer's wave travels that distance, it has completed **one full cycle** and effectively starts over in phase with the wave from the main speaker.

In this case, the correct phase setting would be **0 degrees**, because the two waves are already aligned at the listening position. The same is true if the acoustic center of a perfectly designed subwoofer is physically aligned with the acoustic center of the main speaker. In both scenarios, the peaks and troughs of the two waves coincide, resulting in perfect phase alignment.

The illustration below shows how the two 30 Hz waves overlap precisely at the front of the main speaker when this condition is met.

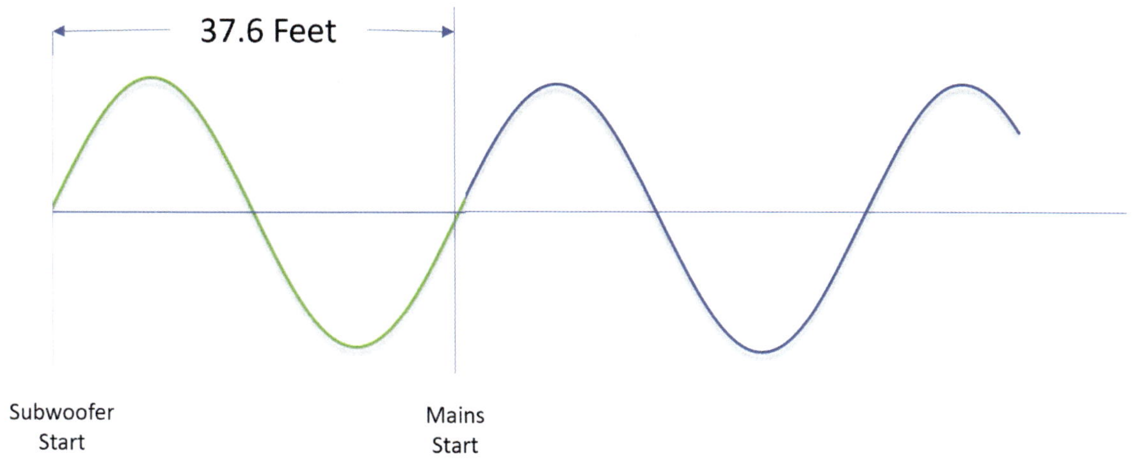

Of course, placing a subwoofer 37.6 feet behind the main speakers is unrealistic, so in any real installation the waves will not be naturally aligned. This means some amount of phase adjustment will always be required, because any placement closer than 37.6 feet produces a phase offset between the subwoofer and the mains.

Let's consider what happens in a few practical cases. First, assume the subwoofer is positioned more than half a wavelength away—that is, farther than 18.8 feet at a 30 Hz crossover frequency. How much phase adjustment is required in this situation?

Before answering, it is important to remember a fundamental limitation: the phase control can only delay the subwoofer's output. It cannot advance the waveform in time. In other words, we can make the subwoofer wait, but we cannot make it arrive early.

In the following illustration, the subwoofer has been moved forward relative to the 37.6-foot reference point, which causes the start of its waveform to shift forward in time. The point where the waveform crosses the x-axis marks 180 degrees of phase rotation. In this example, the subwoofer has been moved forward by less than half a wavelength, meaning the phase offset is less than 180 degrees. As a result, the required phase delay is also less than 180 degrees.

As the subwoofer position approaches 37.6 feet, the required phase adjustment approaches 0 degrees. At 18.8 feet—exactly half a wavelength—the required adjustment is 180 degrees. A position halfway between those distances, at 28.2 feet, corresponds to a 90-degree phase offset.

These specific distances apply only to a **30 Hz crossover**. If the crossover frequency changes—for example, to 40 Hz—the distances will change accordingly. The key takeaway is not the numbers themselves, but the principle: **the half-wavelength distance is where phase polarity inverts.**

Understanding this relationship is essential, because it explains why phase behavior changes predictably as the subwoofer is moved closer to the main speakers—and why the phase control must always be used to compensate for physical placement.

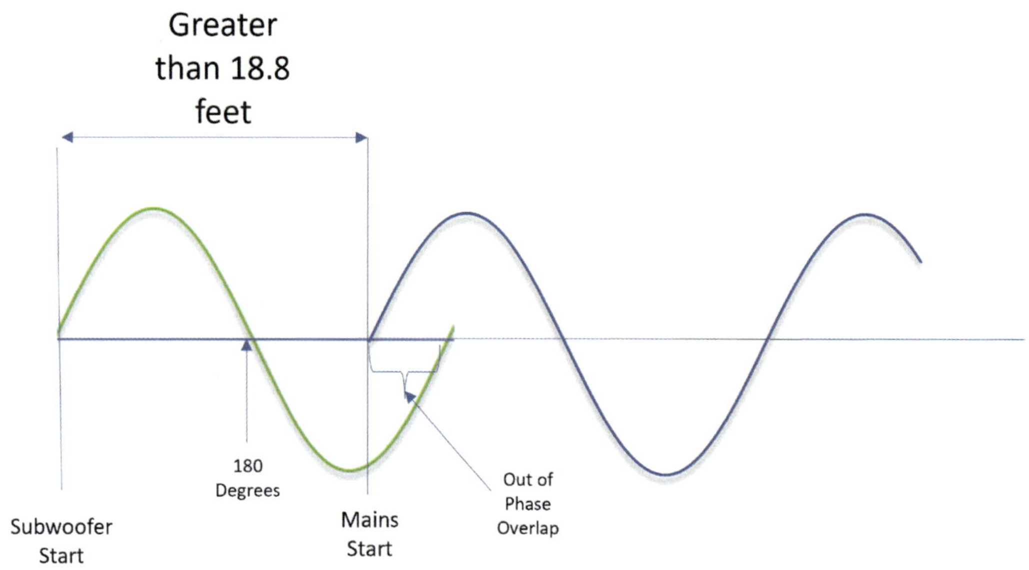

That said, a distance of **18.8 feet** is still impractical in most rooms. In reality, there is a very high likelihood—well over **99% of cases**—that the subwoofer will be positioned **much closer than 18.8 feet** to the main speakers.

In the illustration on the next page, the subwoofer has been moved even closer to the mains. As a result, the waveform has shifted forward further in time, and the **180-degree phase point now lies in front of the main speaker**. In this situation, simply delaying the subwoofer's waveform by less than half a cycle is no longer sufficient. To realign the waves so they sum correctly at the main speaker's position, the phase adjustment must now exceed **180 degrees**.

This is a critical observation: once the subwoofer is positioned closer than half a wavelength at the crossover frequency, achieving proper alignment requires a phase delay **greater than 180 degrees**, because the wave must be delayed far enough to "wrap around" and realign with the main speaker's output.

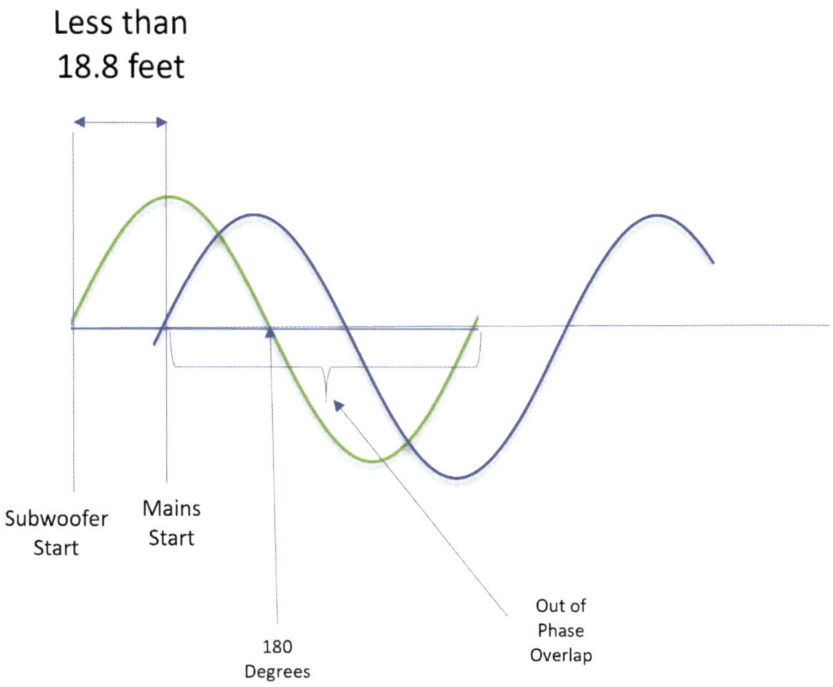

There is a straightforward formula that can be used to estimate the **required phase delay** based on the physical distance between the subwoofer and the main speakers. In this formula, **F** represents the crossover frequency in hertz, and **L** represents the distance between the main speaker and the subwoofer in feet.

Phase (in degrees) = 360 – (0.319 × F × L)

The constant 0.319 accounts for the speed of sound in air and converts distance and frequency into degrees of phase rotation.

The following examples illustrate how the formula is applied:

- **Crossover** = **30 Hz**, distance = **6 ft 3 in (6.25 ft)** Phase = 360 – (0.319 × 30 × 6.25) = **300 degrees**
- **Crossover** = **45 Hz**, distance = **1 ft 4 in (1.33 ft)** Phase = 360 – (0.319 × 45 × 1.33) = **341 degrees**
- **Crossover** = **35 Hz**, distance = **21 ft** Phase = 360 – (0.319 × 35 × 21) = **127 degrees**

These calculations provide a useful **starting point** for setting the phase control. They will not yield a perfect result on their own, but they help place the phase adjustment in the correct range before refining it through measurement or careful listening.

Group Delay

The formula on the previous page works well in an idealized world, but in practice there is an important complication—**group delay**. Technically, group delay is defined as the rate of change of the phase response with respect to frequency, and it is a measurable characteristic of a subwoofer. While that definition may sound abstract, the underlying concept is straightforward. **Group delay is the frequency-dependent time difference between when an input signal arrives and when the woofer actually begins to move.** A theoretically perfect subwoofer would exhibit zero group delay, but in the real world, every subwoofer has some amount of it.

The key phrase here is *frequency dependent.* As frequency decreases, **group delay increases**. Producing higher bass frequencies—such as 80 Hz or 100 Hz—is relatively easy for a subwoofer. As frequencies drop lower, however, maintaining timing accuracy becomes increasingly difficult, both for the signal-processing chain and for the mechanical behavior of the woofer itself. The deeper the frequency, the greater the delay, and the more the subwoofer falls out of time relative to the main speakers.

The group-delay curves below illustrate this behavior for three different subwoofers. In all three cases, group delay rises steadily as frequency decreases from 100 Hz toward 10 Hz. The differences between the designs, however, are substantial. The bottom curve represents a well-behaved subwoofer, maintaining group delay below **10 milliseconds down to 20 Hz**. The middle curve is borderline and may be acceptable in some systems. The top curve is clearly problematic and unsuitable for accurate music reproduction.

As a practical guideline, the **total group delay should be kept below 20 milliseconds**. Once group delay rises above **30–40 milliseconds**, bass reproduction begins to sound slow, detached, and sluggish—undermining both timing and musical coherence.

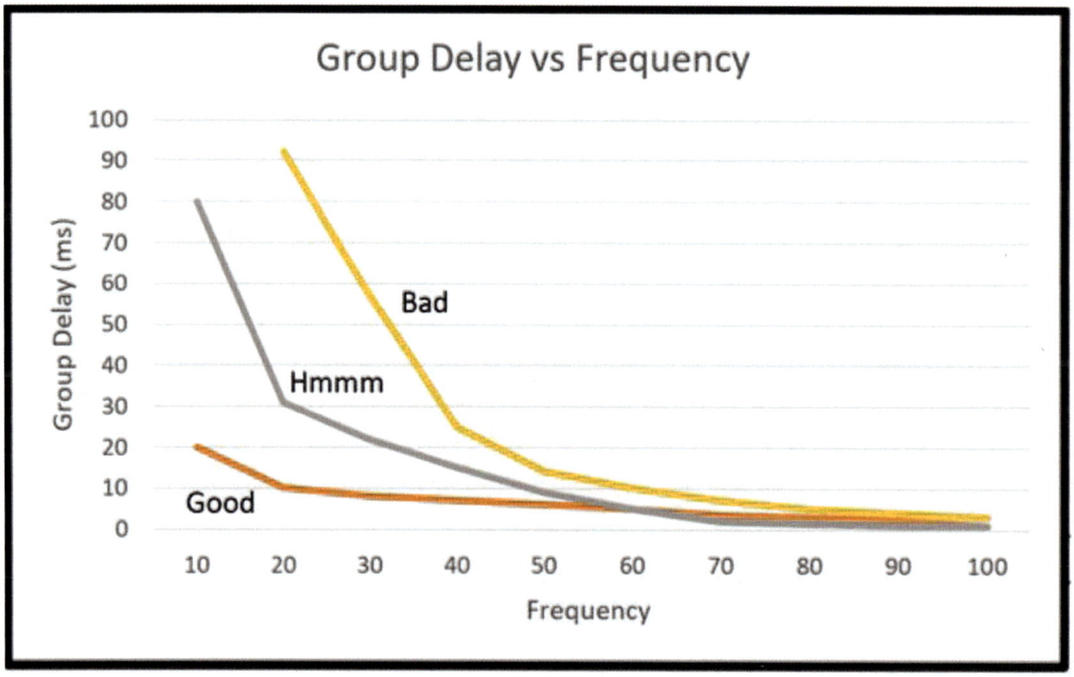

Causes of Group Delay

Several factors contribute to **subwoofer group delay**: the enclosure type, crossover filters, signal processing, and the driver itself. Some of these contributors are more significant than others, but all share one important characteristic—**group delay increases as frequency decreases**. The combined effect of these factors determines the subwoofer's total group delay.

Enclosure Type

The **enclosure design** is one of the largest contributors to group delay. Earlier, I noted that sealed enclosures offer the best chance of success for musical integration, and this is largely due to their group-delay behavior. A sealed enclosure exhibits a **smooth, gradual increase in group delay** as frequency decreases, with a typical value of around **10 milliseconds at 30 Hz**.

By contrast, **ported and passive-radiator designs** exhibit substantially higher group delay. As the operating frequency approaches the enclosure's tuning frequency, group delay rises rapidly. At 30 Hz, values in the range of **20–30 milliseconds** are common. This elevated delay is one reason these designs can sound slower or less controlled when used for music reproduction, even if their frequency response appears favorable.

Crossover Type

The **crossover filter** used to integrate the subwoofer also significantly affects group delay, and this is typically an area where the user has direct control. The most common crossover filter types are **Bessel**, **Butterworth**, and **Linkwitz–Riley**, each of which involves trade-offs between amplitude response, phase behavior, and transient performance.

In general, there is a balance between transient accuracy and amplitude summation:

- **Bessel filters** offer the best transient response and an almost linear phase characteristic, resulting in the lowest group delay.
- **Linkwitz–Riley filters** provide perfect amplitude and phase summation at the crossover frequency, but at the cost of **higher group delay** and poorer transient response.
- **Butterworth filters** fall between the two, offering a compromise in both amplitude and time-domain behavior.

For subwoofer integration, we are primarily concerned with how much **group delay** each filter introduces, rather than how flat the summed response appears on a graph.

Crossover Slope

The **slope of the crossover** also plays a critical role. Increasing the crossover slope will **always increase group delay**. Steeper slopes improve frequency-domain separation but worsen time-domain behavior. The table below shows typical group-delay values at **80 Hz** for various crossover slopes. Keep in mind that as the crossover frequency is lowered, **group delay increases further**, making these effects even more pronounced at subwoofer crossover frequencies.

Crossover Group Delay		
	12 dB/Oct	24 dB/Oct
Bessel	2-4 ms	5-8 ms
Butterworth	3-5 ms	8-12 ms
Linkwitz-Riley	4-6 ms	10-15 ms

Adding **DSP filtering** to correct the frequency response will **increase group delay**. The most common filter used in subwoofer DSP systems is the **IIR (Infinite Impulse Response) bell filter**, which is applied to boost or cut specific frequency regions in order to smooth the amplitude response. While this can improve how the frequency response looks, it always comes with a **time-domain penalty**.

A bell filter is defined by three adjustable parameters: **center frequency**, **gain**, and **Q factor**. Each of these directly affects group delay.

- **Lower center frequencies** result in greater group delay.
- **Boosting gain** introduces more group delay than cutting gain.
- **Higher Q values** (narrower filters) increase group delay significantly.

To understand the impact, consider how changes to a bell filter affect group delay. With a **Q of 1**, applying a bell filter below **40 Hz** will typically add **5 to 12 milliseconds** of group delay when cutting gain. Applying the same filter at a higher frequency—around **60 to 70 Hz**—reduces the added delay to approximately **3 to 5 milliseconds**.

As the **Q factor increases**, group delay rises rapidly. Increasing Q to **4** adds roughly **8 to 18 milliseconds** of delay when the filter is centered below 40 Hz. Increasing Q further to **10** adds approximately **15 to 30 milliseconds** of delay in that same low-frequency range.

Applying a **high-Q boost at 30 Hz** to flatten the frequency response carries a particularly steep cost. Such a filter can add **around 30 milliseconds of group delay** on its own. For this reason, extreme caution is warranted when using DSP to flatten a subwoofer's response. More often than not, the audible cost in timing and articulation outweighs the visual improvement on a graph.

The **driver itself** has the least impact on group delay and is also the aspect over which the user has the least control. We generally assume that the subwoofer designer has made competent choices here. In theory, the ideal driver would be infinitely stiff and massless—but real-world physics does not allow this. Driver mass and material directly affect transient response, as does the surround material. A **lighter, stiffer moving assembly** will respond more quickly. This is a direct consequence of **Newton's second law of motion**. Surround materials also matter: rubber surrounds, for example, tend to absorb energy and recover more slowly, which increases delay.

A hypothetical example helps illustrate how these factors accumulate. Suppose we select a **sealed subwoofer** with a group delay of **10 milliseconds at 30 Hz**. We then implement a **Linkwitz–Riley crossover** at 30 Hz with a **24 dB/octave slope**, which adds **20 milliseconds** of group delay. Next, we apply a bell filter at **35 Hz**, boosting **3 dB** with a **Q of 10** to fill a small dip in the response. This filter adds another **25 milliseconds** of delay.

When combined, the total group delay becomes:

10 ms + 20 ms + 25 ms = 50 milliseconds

At that point, the bass will sound slow, disconnected, and poorly integrated—despite what the frequency-response graph might suggest.

Working With Group Delay

A subwoofer with **high group delay** creates two distinct problems. First, it introduces a **built-in timing delay at the crossover frequency** itself. Second, even if the subwoofer is phase-aligned at one frequency—say **40 Hz**—the increasing group delay at lower frequencies means that by the time the response reaches **30 Hz**, the subwoofer will be delayed further and will no longer be in phase with the main speakers.

This is why, as noted earlier, a **sealed subwoofer** offers the highest probability of successful integration. Sealed designs have the **lowest intrinsic group delay of box designs**, typically **10 milliseconds or less** across the operating range. Ported subwoofers inherently exhibit greater group delay, often in the range of **20 to 30 milliseconds**, and sometimes more. **Passive-radiator designs** generally have even higher group delay than ported enclosures.

The following illustration shows the practical effect of group delay on alignment. From a timing perspective, group delay effectively **adds distance** between the subwoofer and the main speakers. The subwoofer behaves as though it were farther away than its physical location suggests. In other words, the **effective distance** between the subwoofer and the mains is the sum of the measured physical distance, **L**, plus the distance equivalent of the subwoofer's group delay.

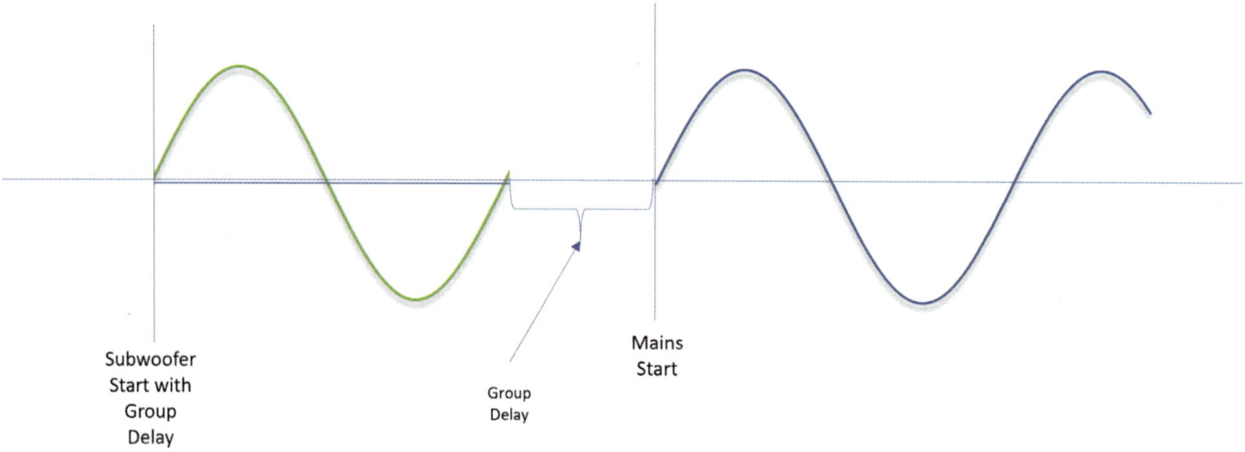

Accurately calculating the correct **phase control setting** requires knowledge of the **group delay of the specific subwoofer** being used—along with any additional delay introduced by the crossover and amplifier if the subwoofer is passive. In some cases, published measurements may be available online that list a subwoofer's group-delay behavior. If that information cannot be found, reasonable assumptions can be made.

As a practical guideline, assume a group delay of **8–10 milliseconds** for a sealed subwoofer and approximately **25 milliseconds** for a ported design. Sound travels at roughly **1.1 feet per millisecond**, which means these delays translate into additional effective distances of about **10 feet** for a sealed subwoofer and **27.5 feet** for a ported subwoofer.

Let's revisit the earlier phase calculations, this time including group delay along with the measured physical distance. In each case, we will calculate phase for both a sealed and a ported subwoofer.

Case 1: 30 Hz crossover, physical distance = 6 ft 3 in (6.25 ft)

- Sealed: 6.25 + 11 = 17.25 ft

  Phase = 360 − (0.319 × 30 × 17.25) = **195 degrees**

- Ported: 6.25 + 27.5 = 33.75 ft

  Phase = 360 − (0.319 × 30 × 33.75) = **37 degrees**

Case 2: 45 Hz crossover, physical distance = 1 ft 4 in (1.25 ft)

- Sealed: 1.25 + 11 = 12.25 ft

  Phase = 360 − (0.319 × 45 × 12.25) = **184 degrees**

- Ported: 1.25 + 27.5 = 28.75 ft

  Phase = 360 − (0.319 × 45 × 28.75) = **−53 degrees**

Case 3: 35 Hz crossover, physical distance = 21 ft

- Sealed: 21 + 11 = 32 ft (≈ 16.1 ft effective at 45 Hz reference)

  Phase = 360 − (0.319 × 45 × 16.1) = **129 degrees**

- Ported: 21 + 27.5 = 38.5 ft

  Phase = 360 − (0.319 × 45 × 38.5) = **−192 degrees**

These examples make the impact of group delay immediately apparent. In many cases, the **effective distance added by group delay exceeds the actual physical distance** between the subwoofer and the main speaker. Earlier, we noted that a physical distance of 18.8 feet (half a wavelength at 30 Hz) was already impractical. With group delay included, we now see effective distances that exceed **a full wavelength**.

A **negative phase value** indicates an effective distance greater than one full wavelength. This is not a desirable situation. When the effective distance exceeds a full wavelength, phase alignment becomes increasingly difficult and less stable across frequency.

Even when the subwoofer's group delay is known, this calculation method provides only a **rough starting estimate** for the phase setting. If group delay is unknown, the assumed values for sealed or ported designs can still place the phase control in the correct general range—but not with precision.

So why perform these calculations at all, if they cannot produce an exact answer? Because understanding the **underlying concept** is essential. To achieve a coherent wave launch from the front of the main speakers, we must understand what the phase control is actually compensating for. Establishing a reasonable initial phase setting is critical. If we do not get close, the result will be audibly poor—and no amount of fine-tuning later will fully correct it!

Determining the Phase Setting by Measuring

Now let's look at what happens when you rotate the **phase control**. Figure 43 demonstrates the combined response at phase settings of **0, 45, 90, 135, 180, 225, 270, 315, and 360 degrees**. The plot also includes a trace for the **correct phase setting**, which in this example was **140 degrees**. The subwoofer crossover in this measurement was set to **30 Hz**.

This is a standard frequency-response plot, with **loudness on the Y-axis** and **frequency on the X-axis**, covering a range from **20 Hz to 100 Hz**.

Figure 43: Subwoofer Plus Mains Varied Phase Values

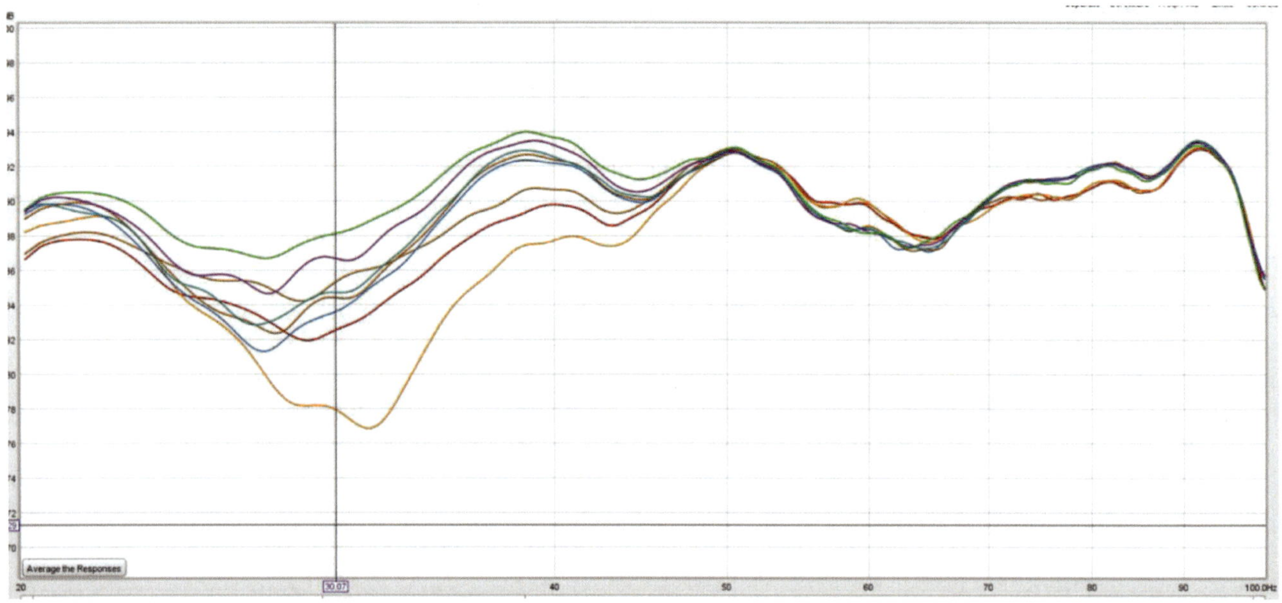

In the graph, notice how dramatically the **perceived subwoofer output changes** as the phase control is rotated through a full cycle. When the subwoofer is close to **180 degrees out of phase** with the main speakers, significant cancellation occurs. In those positions, the system can actually produce **less bass than it did before the subwoofer was added**—clearly not the desired outcome. When the correct phase setting is found, the opposite happens: bass energy is reinforced at the crossover point, and extension improves all the way down to **20 Hz**.

If you are comfortable using **REW** or similar measurement software, this approach will get you to the correct phase setting quickly and with minimal guesswork. Taking measurements at **0, 45, 90, 135, 180, 225, 270, 315, and 360 degrees** typically takes about **30 minutes** and effectively maps the system's behavior. You are simply looking for the response that produces the **maximum output**, because that indicates the subwoofer and main speakers are summing correctly.

There is an even simpler—and somewhat counterintuitive—method for finding the correct phase setting: **look for the worst one**. When the subwoofer is exactly out of phase with the main speakers, a **deep null (cancellation)** will appear at the crossover frequency. This null is unmistakable on a frequency-response plot. Once you locate this point, the subwoofer is known to be exactly **180 degrees out of phase**. All that remains is to shift the phase by 180 degrees, so the subwoofer is brought back into alignment with the mains.

There are three practical ways to do this:

- If the subwoofer includes a **polarity switch**, simply flip the switch.
- If you are using a **passive subwoofer**, reverse the positive and negative speaker connections at the subwoofer terminals.
- Take the current phase setting in degrees and **subtract 180 degrees**. Use the resulting value as the new phase setting.

The frequency-response graph in Figure 44 illustrates this method clearly. In this example, a deep null appears when the phase control is set to **320 degrees**. Flipping polarity shifts the phase to **140 degrees**, which produces the strongest and most coherent bass response.

Figure 44 Frequency-response Graph

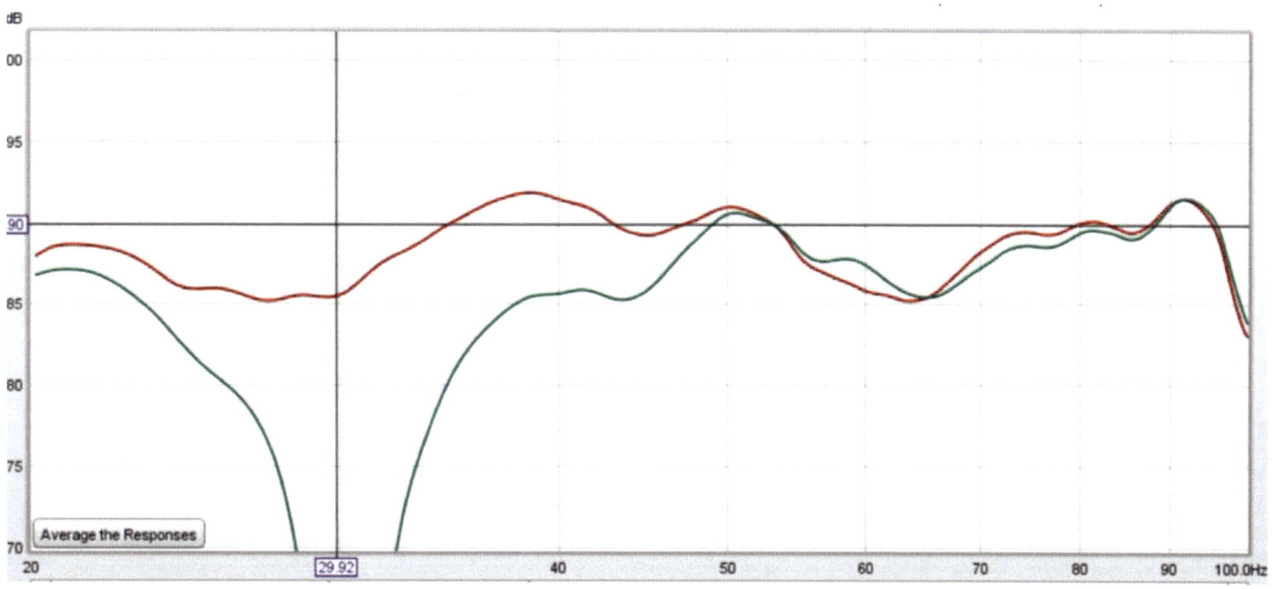

Finding the Correct Phase Without Measuring

Math and measurements aren't for everyone—and this is likely the section many of you have been waiting for. So how do we find the correct phase setting **without measuring**? There are two practical approaches. One uses a simple **SPL meter and a test tone**. The other relies entirely on **listening**. Both methods are based on the same principle: when the subwoofer and main speakers are **in phase**, their outputs add together and the sound is louder; when they are **out of phase**, cancellation occurs and the sound becomes noticeably quieter.

Determine the Correct Polarity

Begin by finding the correct **polarity**—either 0 or 180 degrees. Play music with strong, deep bass (for example, *Poem of Chinese Drum*), or use a warble tone at the crossover frequency. Compare the two polarity settings and determine which one is **louder**. You can make this judgment by ear or with an SPL meter. The louder setting indicates the correct polarity.

Once the correct polarity has been identified, deliberately switch to the **wrong polarity**—the quieter one. This is the setting that will produce the deepest cancellation and allow us to find the correct phase position.

Next, play a test tone at the **crossover frequency**. If you do not have a test-tone CD, an online source will work just as well. In the example discussed earlier, the crossover frequency was **30 Hz**, so a 30 Hz tone was used.

Method 1: Using an SPL Meter

With the system playing the test tone and the polarity set incorrectly, slowly rotate the phase control while watching the SPL meter. Find the position where the meter shows the **lowest possible reading**. This point represents maximum cancellation. Once you have found it, flip the polarity **180 degrees**. The subwoofer is now correctly phased with the main speakers.

Method 2: Using Your Ears

This method takes advantage of how our brains localize sound, even at very low frequencies. Start by playing *Poem of Chinese Drum* **with no subwoofers active**. The drum should sound perfectly centered.

Now turn on the **left subwoofer only**. As you rotate the phase control, listen carefully. At a certain point, the drum will begin to **pull toward the right**. This happens because the left subwoofer is canceling bass energy on the left side. Continue adjusting the phase until the drum is pulled as far right as it will go—this is the point of **maximum phase cancellation**.

Once you reach that point, flip the polarity **180 degrees**. The drum should snap back to the center and sound powerful and well defined. Repeat this same process for the **right subwoofer**.

Wrapping Up Phase

This section is admittedly demanding, but if you have stayed with it, you should now have a clear and repeatable framework for setting subwoofer phase without guesswork. Before moving on, it is worth reinforcing a few key principles that underpin everything we have just covered:

1. **The sound wave produced by the subwoofer must be in phase with the wave launched by the main speaker.** When this condition is met, the two sources sum coherently and reinforce one another.
2. **Subwoofers exhibit group delay**, which can add a small or significant amount of effective distance between the subwoofer and the main speakers.
3. **Phase can be expressed in three equivalent units**—distance, time, or degrees. These units are directly related, and phase expressed in one can be converted to the others.
4. **Polarity is phase inversion.** It is either 0 degrees or 180 degrees and simply flips the waveform upside down.
5. **The phase control on a subwoofer can only delay the sound wave**; it cannot advance it.
6. **The correct phase setting depends on the crossover frequency**, because phase expressed in degrees is frequency-dependent.
7. **The phase position that produces a deep null at the crossover frequency represents exactly 180 degrees of phase misalignment.**
8. **The correct phase setting is the one that maximizes combined output**, allowing the subwoofer and main speakers to sum cleanly and extend bass response without cancellation.

With these principles in mind, the phase control becomes far less mysterious. It is no longer a matter of guessing, but of understanding what the control is compensating for—and how to arrive at the correct setting reliably.

Something Extra: EQ

Many subwoofers include the ability to **equalize (EQ)** their frequency response. Several types of EQ filters exist, but the one we will focus on here is the **bell filter**. It is named for its characteristic bell-shaped curve.

A bell filter is particularly effective at **reducing localized peaks** in the frequency response. For example, earlier we examined a frequency plot that showed a pronounced peak at **55 Hz** and noted that it could be reduced if

DSP were available. A bell filter is well suited to this task because the peak itself closely resembles a bell-shaped curve. By matching the filter to the shape of the peak, we can reduce its level without unnecessarily affecting surrounding frequencies.

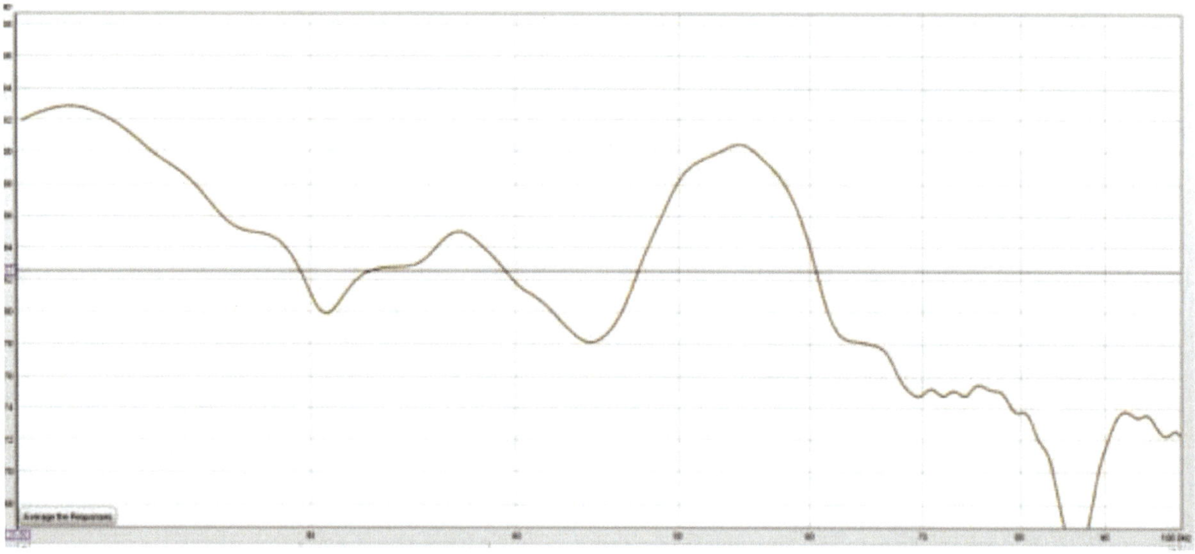

A **bell filter** is defined by three adjustable parameters: **frequency**, **gain**, and **Q**. These values can be entered either through a set of physical controls on the subwoofer or via DSP software.

Frequency specifies the center of the peak you want to address. Referring to the example graph, the peak occurs at approximately **55 Hz**, so that is the frequency value we would enter.

Gain determines how much the level at that frequency is boosted or reduced. In this case, the peak rises slightly above **90 dB**. If we draw a vertical line downward from the peak at 55 Hz to the level we want to achieve—about **80 dB**—we can see that the peak needs to be reduced by roughly **10 dB**. The appropriate gain setting would therefore be **–10 dB**. Be sure to include the minus sign, as this is a cut rather than a boost.

The **Q factor** controls the width of the bell curve. A low Q produces a broad adjustment, while a high Q produces a narrow one. Q is defined as the **center frequency divided by the bandwidth at –3 dB** from the peak.

To estimate Q from the graph, draw a horizontal line **3 dB below the peak.** Since the peak is around 90 dB, the –3 dB line would be at **87 dB.** Note the frequencies where the response crosses this line on either side of the peak. In this example, those points occur at approximately **49 Hz** and **58 Hz**. Subtracting these values gives a bandwidth of **9 Hz**. Dividing the center frequency by this width yields:

$55 \div 9 \approx \mathbf{6}$

In practice, rounding up slightly to **Q ≈ 8–10** is reasonable for a narrow peak like this. For illustration, we will use **Q = 10** as a starting point.

At this stage, we have our initial bell-filter settings:

- **Frequency:** 55 Hz
- **Gain:** –10 dB
- **Q:** 10

These values should be entered as a starting point. After applying the filter, take another measurement to see how effectively the peak has been reduced. From there, make small adjustments as needed. Just as important, **listen carefully**. A flatter-looking response does not automatically mean better sound.

There are several important cautions when using EQ:

1. **Avoid boosting whenever possible.** Every 3 dB of boost requires **twice the amplifier power** at that frequency. In many cases, boosting will not successfully fill a dip in the response. If boosting is unavoidable, keep it below **3 dB**.
2. **Be cautious with high-Q filters.** Narrow, high-Q adjustments can have unintended negative effects on transient response and overall sound quality. If a peak is extremely narrow, repositioning the subwoofer is often a better solution than aggressive EQ.
3. **Use as few filters as possible.** Proper placement should do most of the work. In the example shown earlier, a flat response from **20 Hz to 80 Hz** was achieved using only **two EQ filters per subwoofer**.

Remember that every EQ filter increases **group delay**. Excessive filtering can result in bass that sounds slow, thick, or disconnected. EQ should be used as a finishing tool—not as a substitute for correct placement and alignment.

Integrating a Subwoofer

At this point, your subwoofers should be placed in their **best candidate locations**, and you should have established solid starting values for **crossover frequency, gain, and phase**. Now it is time to focus on **grid alignment**.

Begin by matching the **toe-in** of the subwoofers to that of the main speakers. Using a Digi-Pas level, set the **rake angle to 0.1 degrees** and the **azimuth to 0.000 degrees**. If your system includes large bass towers integrated into the speaker structure, align the rake, azimuth, and toe-in of the bass section to match the main speaker exactly.

This process should feel familiar. It is the same approach used earlier when positioning the right speaker: measure carefully, make deliberate adjustments, and establish a precise starting point. The same discipline applies here. A smooth frequency response alone is not sufficient. The **wave launch from the subwoofer must be time-aligned** with the wave launched from the main speakers.

Like the main speakers, a subwoofer has **six degrees of freedom**, and each one must be addressed to achieve seamless integration.

A useful way to visualize this is to imagine the bass energy from the main speaker and the subwoofer as **two expanding bubbles** moving toward the listener. For proper integration, the **grid lines within those bubbles must align**—the bubbles should move in the same direction and expand at the same rate. When this happens, the subwoofer ceases to sound like a separate source and instead becomes an extension of the main speaker.

The primary reference tracks used for subwoofer alignment are:

- *Cosmo... Old Friend*
- *L'Italiana in Algeri*
- *Poem of Chinese Drum*
- Shostakovich, *Symphony No. 8*, movements III and IV
- *Till Tomorrow*
- Bass and Drum Intro

These selections provide the timing cues, spatial information, and low-frequency articulation required to fine-tune alignment across all six degrees of freedom.

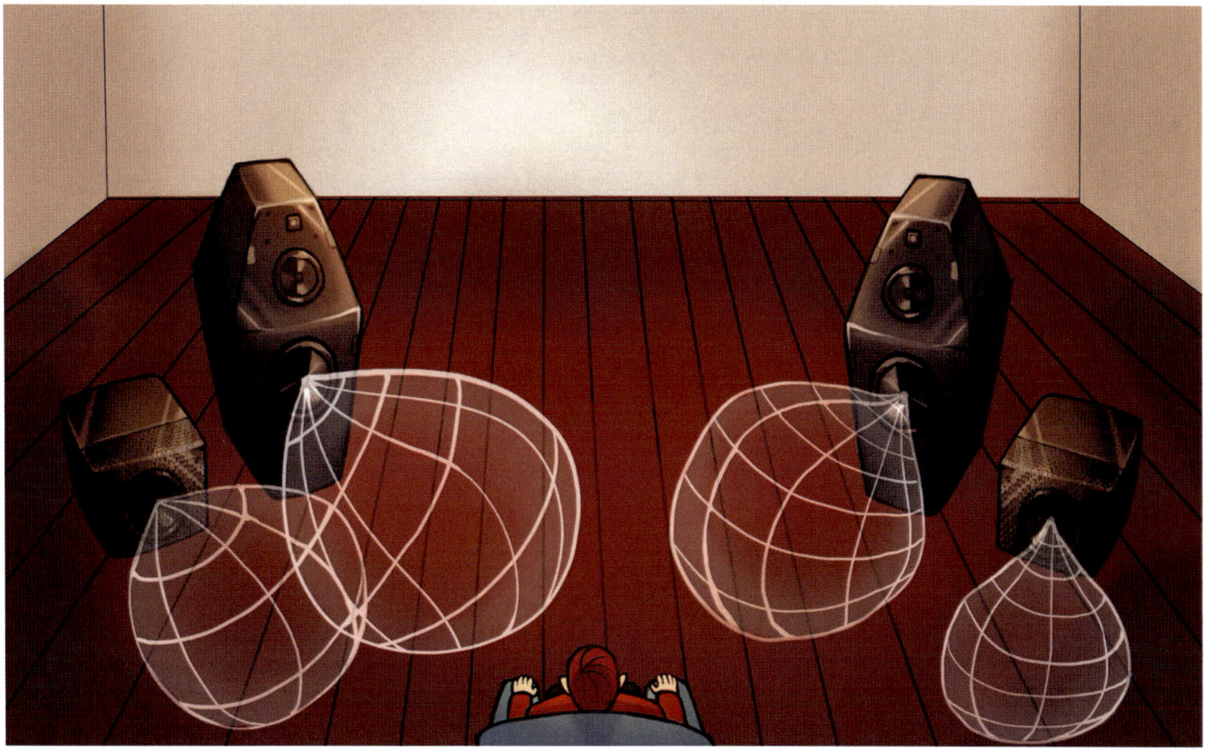

In the image above, the speaker on the right has a subwoofer placed beside it that is **not integrated**. The illustration shows two distinct sound "bubbles," which means the subwoofer is audible as a separate source—an undesirable result. Unfortunately, this is how many subwoofers are perceived, and it explains why so many listeners abandon them altogether.

Now look at the image on the left. Here, the sound bubbles from the main speaker and the subwoofer **overlap**, and their grid lines are aligned. However, because the main speaker's bubble is larger and reaches the listener first (because it's physically closer), the subwoofer's timing has been aligned so that both sources expand coherently. When the grid lines line up, the subwoofer no longer calls attention to itself; it simply reinforces the foundation of the main speaker.

The grid lines are aligned by manipulating the **six degrees of freedom**, exactly as we did with the main speakers:

- **Height**: Adjusting height increases bass articulation. The bass rises to meet the rest of the sound instead of sounding like it is stuck on the floor. Adjust height in **1 mm increments** until the bass sounds lifted and connected.
- **Rake**: Rake controls how bass energy is released into the room. Start with a rake angle of **0.1 degrees backward**, then continue adjusting until the bass flows over you rather than pooling at the speaker.
- **Front-to-back position**: Moving the subwoofer forward or backward adjusts timing and aligns the bass wave with the main speaker.

- **Lateral position**: Side-to-side movement increases or decreases coupling with the main speaker. Small lateral adjustments can add clarity or remove blur.
- **Azimuth**: This becomes important when using tall bass towers positioned next to the main speakers.
- **Toe-in**: Toe-in affects how bass pressure is perceived and how well it centers.

To make these adjustments, you will need a way to change **height, rake, and azimuth**. The ideal solution is a **threaded, continuously adjustable foot**. One effective option is a spiked foot with approximately one inch of adjustable thread. In practice, most subwoofers benefit from **additional height**, and adjustable feet provide a usable starting point that can be fine-tuned upward.

If adjustable feet are not available, you can make effective temporary adjustments using simple materials. Cut several **3" × 3" MDF squares**—precision is not critical—and place them under the subwoofer's feet. Having pieces in **1/8", 1/4", 1/2", and 3/4" thicknesses** allows you to adjust height in **1/8-inch increments** by stacking them. Once you hear the improvement this makes, you will likely want to invest in proper adjustable feet.

Finally, pay close attention to **sound pressure**. This cue is essential when integrating subwoofers and even more important than it was with the main speakers. As frequency drops below **60 Hz**, our perception transitions from hearing sound to **feeling** it. This is precisely the region where subwoofers operate.

You can use pressure cues to align each subwoofer individually, but when matching the left and right subwoofers, your body becomes an invaluable reference. By noticing **where** the pressure wave strikes you and **from which direction**, you can tell how the right subwoofer must move to match the left. When pressure feels even, centered, and forward-moving, the subwoofers are beginning to act as a single, coherent foundation rather than two separate sources.

Tying It All Together: A Hypothetical Setup

At this stage, I have measured my main speakers and confirmed that they play flat down to **35 Hz**, after which they begin to roll off. To extend the low end, I have chosen a **pair of sealed, passive subwoofers** designed to integrate cleanly with the mains. I measured the longest distance to any candidate placement location and ordered an inexpensive set of interconnects long enough to reach all possibilities.

Once the subwoofers arrived, I placed them in the room and connected them. I work with **one subwoofer at a time**, since they will ultimately be placed symmetrically. I always integrate the **left subwoofer first**, then duplicate the result on the right.

Initial Placement and Phase Estimates

I identified three candidate locations and placed the **left subwoofer** in each position in turn. For each location, I measured the distance from the **front of the main speaker** to the subwoofer's woofer and added an estimated group delay. I then converted that total distance into an approximate phase setting in degrees—close enough to get started.

The candidate positions were:

- **Position 1:** Next to the main speaker (distance = 0 ft)
- **Position 2:** Just behind and inside the main speaker (distance = 2.5 ft)
- **Position 3:** Near the front corner (distance = 6.3 ft)

The crossover point is approximately **33 Hz**. Because I am using a high-quality sealed subwoofer with an external active crossover and no DSP, I assume a low group delay of **5–6 ms**.

Using the phase formula:

- **Position 1:** $360 - (0.319 \times 33 \times (0 + 5.5)) \approx$ **302°**
- **Position 2:** $360 - (0.319 \times 33 \times (2.5 + 5.5)) \approx$ **276°**
- **Position 3:** $360 - (0.319 \times 33 \times (6.3 + 5.5)) \approx$ **236°**

These values are **starting points**, nothing more.

Evaluating Each Position

At the first position, I set the crossover near **33 Hz**, adjusted gain to roughly match the mains, and set phase to **302°**. I then measured the subwoofer's response and listened for usable bass between **20 and 35 Hz**, comparing the sound with and without the subwoofer engaged.

I repeated this process for the second and third positions, adjusting phase to **276°** and **236°**, respectively.

At this point, an obvious limitation becomes clear: **phase knobs are not precise**. Hitting exact degree values is unrealistic. At **35 Hz**, one degree of phase corresponds to roughly **1.14 inches** of physical distance. Even perfect knob placement still leaves you within a wide margin. Final alignment requires **physical movement plus small phase adjustments**.

I compared the three frequency plots from **20 to 80 Hz**, looking for:

- Flat response below **35 Hz**
- No large peaks above the crossover (especially **50–60 Hz**)

I selected the best-performing location and explored that area further. When possible, I laid out a **small grid (nine positions)** and measured each one, then moved the subwoofer to the best spot. I noted the final distance, recalculated phase, adjusted the knob, matched toe-in to the mains, and set **rake to 0.1°** and **azimuth to 0.000°**.

Listening and Fine Adjustment

Now the real work begins: **listening**.

I cycled through the setup tracks, focusing on short sections rather than entire pieces:

1. **"Cosmo… Old Friend"** I listen to the bass drum:
    - » Is it centered?
    - » Is it smeared?
    - » Where does the lowest energy appear?
    - » How much overhang is there?
2. ***Shostakovich Symphony No. 8*** (Movements III–IV)
    - » End of the third movement: timpani clarity
    - » Beginning of the fourth: the three "bomb blasts"
    - » Transient and bass wave must feel connected
    - » On the third blast, the bass should fully exhaust into the room
3. **L'Italiana in Algeri**
    - » Pizzicato bass articulation
    - » No bass overhang

» No chuffing from cellos

» Oboe solo must sound as good—or better—than without the subwoofer

4. **"Poem of Chinese Drum"**

» Drum must be centered

» Strong chest impact

» Clear transient followed by body, not detached boom

5. **"Bass and Drum Intro"**

» Used last, primarily to adjust height

Correcting the Big Problems

Two issues almost always appear first:

1. Bass Ground Fog

Using *Bass and Drum Intro* (1:07–1:13), listen to the low E string. If the bass sounds disconnected and stuck near the floor, raise the subwoofer:

- Spikes: quarter-turn increments
- Shims: start with **¼" MDF**

Raise until the bass connects to the instrument.

2. Bass Overhang

Return to *Cosmo… Old Friend*. If the drum sounds like **boom–boooom**, move the subwoofer **forward** in small increments until the decay tightens.

Rake and Transient Alignment

In Shostakovich 8, the bass wave must not arrive before the transient. Adjust rake in **0.01–0.02° increments**, tilting back until the bass exhausts naturally into the room—like saying "boom" while fully exhaling.

Compensate rake changes with small forward bumps if needed. Recheck height and overhang.

Lateral Alignment and Imaging

Center the drum image using lateral movement:

- If the image pulls left → bump right
- If it pulls right → bump left

Despite common belief, **bass below 30 Hz absolutely affects stereo imaging**. You can easily pull images with a subwoofer crossed at 30 Hz—even with a 24 dB/oct slope.

Final Phase Fine-Tuning

In *L'Italiana in Algeri*, listen closely to the oboe solo. This is where **phase is finalized.**

Make small phase adjustments (e.g., from 26° down to 15–20°), listening for:

- Pure vibrato
- No quacking or squawking
- Sound that floats into the room

After phase changes, recheck *Cosmo... Old Friend* and trim overhang if needed.

Integrating the Right Subwoofer

Once the left subwoofer is correct, repeat the entire process for the right—but with one added constraint: **they must work together**.

Match all physical parameters exactly: height, rake, azimuth, toe-in, and front-to-back plane.

Use *Till Tomorrow* by Yello. The bass pulse should sound like **one centered source**:

- Right arrives first → bump right subwoofer back
- Left arrives first → bump right subwoofer forward

Adjust until the pulse locks in.

Use **pressure cues**. Bass should hit straight ahead, evenly. If pressure angles to one side, adjust lateral position and toe-in.

Finally, turn the left subwoofer off and confirm the right alone behaves properly. Then listen with both on, then with both turned off. If bass without subs sounds faster or clearer, return and refine further.

This process rewards patience. Have Fun!

Summary: Completing the Musical Foundation

In Chapter 10, we turned our attention to the lowest frequencies of the system and reconsidered the role of subwoofers within a properly aligned stereo setup. Rather than viewing a subwoofer as a device that simply adds more bass, this chapter reframed its purpose: a subwoofer extends the musical foundation already established by the main speakers.

Because low frequencies interact strongly with the room, subwoofer integration must only occur after the main speakers and listening position have been fully optimized. Attempting to integrate subwoofers before this work is complete typically magnifies existing problems—masking timing errors, exaggerating room modes, and blurring the spatial coherence achieved through careful speaker alignment.

This chapter explained how subwoofers should be introduced so that they reinforce, rather than disturb, the timing relationships and tonal balance already present in the system. Proper integration requires attention to placement, phase alignment, crossover behavior, and level matching so that the subwoofer works in concert with the main speakers rather than competing with them.

When correctly integrated, a subwoofer should never call attention to itself. Instead, it disappears as a source and allows the system to reproduce the full scale and weight of the musical performance. Bass instruments gain natural body and authority, the soundstage acquires greater dimensional stability, and the entire presentation becomes more grounded and believable.

In this way, the subwoofer does not merely add bass—it completes the musical structure that the main speakers have already revealed.

CHAPTER 11

ROOM ACOUSTICS — WORKING WITH THE SPACE

The primary purpose of this guide is to describe **how to position loudspeakers**. The placement principles presented throughout the book are intended to show how to work *with* the room—optimizing sound by minimizing negative room interactions as much as possible. However, even after speaker placement has been fully optimized, some undesirable aspects of the sound may remain. When that happens, it can be appropriate to consider **room adjustments**, which fall under the broader category of **acoustics**.

This chapter on acoustics is meant to **get you started**. It is not intended to be the final word on acoustically treating a room. Its most important purpose is to give you enough understanding to **avoid common traps** that result in frustration and wasted money. A prevailing belief in some audiophile circles is that great sound requires a dedicated room designed by a professional acoustician and treated with large quantities of absorbers and diffusers. Based on experience setting up hundreds of rooms, this is simply not true. You do not need to fill a room with panels to achieve excellent sound.

Acoustic treatment is often applied with the goal of flattening frequency response and controlling reverberation. What room treatments **cannot** do is fix timing problems. What they *can* do is absorb large amounts of acoustic energy—and if applied indiscriminately, this can strip the music of life and engagement. There are many cases where a homeowner hires an acoustician, treats the room extensively, and later removes or reworks much of the treatment because the result is unsatisfying. This chapter is written to help you avoid that experience.

This is not to say that a dedicated room offers no benefits. On the contrary, its greatest advantage is **silence**—freedom from refrigerator hum, traffic noise, HVAC systems, and other background sounds. Lowering the ambient noise floor allows greater musical nuance and expression to emerge. A dedicated room also provides maximum flexibility in speaker and listener placement, offering far more freedom to search for the optimal geometry. That said, a poorly designed dedicated room can easily create new problems that then require correction.

Why Rooms Create Problems

If a dedicated room is so beneficial, why does it still present challenges? The answer is simple: **almost all of the sound produced by the speakers remains trapped inside the space**. Loudspeakers do not project sound only toward the listener—they radiate sound in all directions. These sound waves reflect off the walls, floor, and ceiling, bouncing repeatedly while gradually losing energy. What ultimately reaches your ears is not just the direct sound from the speakers, but a complex mixture of direct sound and reflections from multiple surfaces. At any given moment, what you hear is the sum of all these overlapping waves. When waves combine, they interfere with one another, producing irregularities in the frequency response and time domain.

The scientific study of sound behavior is called **acoustics**. In home audio, we are concerned with a subset known as **small-room acoustics**. Research in this field has identified two primary acoustic problems that dominate sound behavior in listening rooms:

1. **Room modes**
2. **Speaker Boundary Interference Response (SBIR)**, also known as the **Allison effect**

We will examine both.

Room Modes

Because a room is a closed container, it supports standing waves known as **modes**. A mode is a region in the room that exhibits elevated sound pressure at specific frequencies. In formal mathematical terms, a mode is an *eigenfunction*, and the frequency at which it occurs is an *eigenvalue*. Each mode has a specific spatial distribution determined by the room's dimensions and shape.

In a rectangular room, modal behavior follows predictable patterns. Along a given dimension, the pressure distribution is described by a cosine function:

$$\cos(n\pi X / L)$$

where:

- n is an integer representing the mode number,
- L is the length of the room dimension, and
- X is the position along that dimension.

The first-order mode ($n = 1$) produces maximum pressure at the boundaries—walls, floor, and ceiling. This leads to a well-known practical rule: **placing speakers or the listening position close to walls exaggerates certain bass frequencies**, producing boomy, uneven bass. Anyone who has experimented with speaker placement has discovered this intuitively.

An Important Aside

All enclosed spaces have modes. This is a consequence of the physics governing wave behavior in bounded volumes. It is sometimes claimed that round rooms or splayed walls eliminate room modes. This is false. **All rooms have modes.** What changes with room shape is how those modes are distributed spatially. Rectangular rooms have the advantage of predictability—we know exactly where high-pressure regions will occur.

The existence of room modes cannot be avoided, but their effects can be managed.

Practical Observations About Room Modes

Several generalizations are worth keeping in mind:

- The **greater the distance between two opposing boundaries**, the **lower the frequency of the first mode**. At a spacing of **28.5 feet**, the first modal frequency drops below **20 Hz**. If designing a room, making one dimension close to 28.5 feet can be advantageous.
- Each room dimension—length, width, and height—produces its own modal series. Rooms with **similar dimensions** tend to pile multiple modes on the same frequencies, creating large peaks. Avoid rooms where dimensions are close to equal.
- **All modes exhibit high pressure at boundaries.** Keep both speakers and the listening position away from walls when possible.
- **Corners** are ideal locations for acoustic treatment, since all modes exhibit high pressure where boundaries intersect.
- In **L-shaped rooms**, strong modal activity occurs at the inside corner of the "L," making this another area where treatment can be effective.

Speaker Boundary Interference Response (SBIR)

Before addressing SBIR directly, recall a basic property of waves: **superposition.** Waves add together. When two waves arrive at your ear in phase, their amplitudes sum, increasing loudness. When they arrive out of phase, they partially or completely cancel. This interaction creates peaks and nulls in the frequency response.

SBIR—also known as the **Allison effect**, named after Roy F. Allison—occurs when a loudspeaker is placed at a distance *d* from a boundary. Sound radiates directly toward the listener and simultaneously toward the wall. The reflected wave returns to the listener delayed in time. At certain frequencies, this delay causes destructive interference, producing deep nulls in the frequency response.

The frequency of the first SBIR null is given by:

$$f = 1130 / (2L)$$

where *L* is the path-length difference (in feet) between the direct and reflected sound. These nulls are easily visible in measurements taken with tools such as REW.

One important insight from this formula is that **as the speaker moves farther from the wall, the SBIR null shifts to a lower frequency**. This is why speaker distance from boundaries is such a critical variable in low-frequency behavior.

An Example: How SBIR Creates Nulls

An example helps clarify how SBIR works in practice. Suppose we place our speakers **3 feet from the front wall** and want to know where the first SBIR null will occur. Recall that in the SBIR formula, **L** represents the **path-length difference** between the direct sound and the reflected sound. Because the sound must travel to the wall and then back to the listener, this path-length difference can be approximated as **twice the speaker's distance from the wall**.

In this case, the path-length difference is approximately **6 feet**. Plugging that into the formula:

$$f = 1130 / 2L = 1130 / (2 \times 6) \approx 94 \text{ Hz}$$

So, placing the speaker 3 feet from the wall produces a deep null at approximately **94 Hz**.

If we want that null to fall **below the audible range**, we can ask how far from the wall the speaker must be so the null drops below **20 Hz**. Rearranging the formula:

$$L = 1130 / 2F = 1130 / (2 \times 20) \approx 28.25 \text{ ft}$$

This value represents the **round-trip path-length difference**, so the actual speaker distance from the wall would be roughly **half of that**, or about **14 feet**. In other words, to eliminate an audible SBIR null caused by the front wall alone, a speaker would need to be placed roughly **14 feet from that wall**—clearly impractical in most rooms.

Why Rooms Create Multiple Nulls

A rectangular room has **four walls, a floor, and a ceiling**, which means there are **six primary boundaries** that can generate SBIR effects. As a result, the bass region will typically contain **multiple prominent nulls**. Across the full audio spectrum, these interactions produce additional ripples known as **comb filtering**.

There are two broad strategies for dealing with this reality:

1. **Empirical placement** of speakers and listening position so that reflections and modes average out into a smoother response.
2. **Acoustic treatment**, which attempts to absorb, diffuse, or redirect reflections—particularly at early reflection points.

This book focuses primarily on the **first approach**. Acoustic treatment supports the second.

How Sound Interacts with Boundaries

In small-room acoustics, we are primarily concerned with **room modes** and **SBIR**, both of which arise from sound interacting with boundaries. When a sound wave encounters a boundary, one of three things happens:

- The wave is **transmitted** through the boundary
- The wave is **absorbed**
- The wave is **reflected**

Transmission

If sound is transmitted, energy is leaking out of the room—and outside noise can leak in. This is generally undesirable in a listening room. Transmission varies by material and frequency. For example, **glass is highly transmissive at low frequencies** but very reflective at high frequencies. A room with large windows on one side can therefore exhibit uneven left–right bass balance.

Absorption

Absorption reduces the energy of a sound wave by converting acoustic motion into heat. There are two primary mechanisms:

1. **Porous absorption**, where air motion is slowed as it passes through a fibrous or porous material. This is the most common form of acoustic treatment and is typically sold as acoustic "panels." These are usually made from semi-rigid fiberglass and covered with acoustically transparent fabric.

   Absorption effectiveness depends heavily on thickness:

   - 1" panel (mounted to a wall): effective down to ~2000 Hz
   - 2" panel: effective down to ~700 Hz
   - Much thicker panels are required to absorb below 100 Hz—often **on the order of 24 inches**

   As a result, porous absorbers are **ineffective for fixing bass problems** below 80 Hz.

2. **Pressure-based absorption**, used for low frequencies. The most common example is a **damped membrane absorber**—a flexible panel mounted over an airtight cavity. These must be **precisely designed** to target a specific frequency and placed where pressure for that frequency is high. If built or positioned incorrectly, they simply will not work.

Reflection and Diffusion

Hard, flat surfaces reflect sound much like a mirror reflects light. Sometimes this reflection is undesirable but absorbing it entirely can make the room sound dead and lifeless. Instead, we can **diffuse** the reflection—preserving energy while breaking up directional coherence.

Some important rules regarding diffusers:

1. **Diffusers do not work at low frequencies.** A diffuser that operates down to 500 Hz is already considered low-frequency capable. Diffusers cannot address room modes or bass issues.
2. **Distance matters.** The listener and/or speaker must be several wavelengths away. For example, an 800 Hz diffuser (wavelength ≈ 1.4 ft) requires roughly **3–5 feet** of distance to function properly.
3. **Moderation is critical.** Too much diffusion leads to unfocused, chaotic sound.
4. **Not all "diffusers" are real diffusers.** True mathematical diffusers include **quadratic residue diffusers (QRDs)** and **poly-cylindrical diffusers**. Many products merely scatter sound.

Where Acoustic Treatment Often Goes Wrong

The most common recommendation is to place treatment at **first reflection points**. That advice is not wrong—but it is incomplete.

Here's the problem: **you cannot know the first reflection points until the speakers and listening position are finalized**.

Many acousticians treat the room as a static object and assume speakers can be placed "somewhere up front." This often gives rise to panels being installed in locations that later interfere with finding the best speaker position.

Consider this thought experiment. Absorbers and diffusers are installed along the side walls in an even, visually pleasing pattern. You begin speaker setup and find a good spot. Then you move the speaker forward slightly and suddenly the bass growl is spectacular—but the midrange collapses. You move again and find balance—but the magic is gone. Eventually, you settle for a "pretty good" spot.

What you didn't notice is that the **sidewall reflection point moved** from a neutral surface to a treated surface as you moved the speaker. Without treatment in place, the truly optimal position would have revealed itself clearly.

This is why **random or premature treatment can make setup harder**, not easier.

Where Treatment *Can* Be Placed Safely

The **four corners of the room** are the only locations where treatment can be placed independently of speaker position. All modes exhibit high pressure at boundaries, and boundaries intersect in corners. If you plan to use treatment before setup, start here—with **thick corner absorbers**, ideally **16 inches or thicker** if meaningful bass control is desired.

Once the speakers are placed, acoustic treatment can be evaluated and applied deliberately, rather than guessed at. At this stage, we can experiment with limited panel placement to determine whether it improves the musical experience. The most logical place to begin is at the **points of first reflection**.

So how do we find those points?

The **sidewall and ceiling first reflections** are the easiest to locate empirically. Sit in the listening position while another person moves a mirror along the wall or ceiling. When you can see the speaker's tweeter reflected in the mirror from your seated position, you have found a **first reflection point**. This method is simple, fast, and surprisingly accurate.

There is also a mathematical approach known as **ray tracing**, which calculates the path of direct and reflected sound waves based on geometry. Several online tools can assist with this. An example of such a ray-tracing diagram is shown below.

In the next graph, the corner of the room is shown with two points: the point near the corner represents the **speaker**, and the point nearer the center represents the **listener**. Several important observations emerge from this visualization:

- The **front-wall reflection point** is often located **inside the speaker's position**, not directly behind it. It is common to see panels centered on the front wall that completely miss the speaker's actual first reflection point.
- The **sidewall reflection point** is typically **closer to the speaker than expected**, not midway between the speaker and the listener.
- The **corner produces a double reflection**, where energy reflects off two boundaries in quick succession. This reinforces why corners are such effective locations for acoustic absorption.

These observations underscore an important principle: **acoustic treatment should follow speaker placement, not precede it**. When panels are placed after the speakers are optimized, treatment can support clarity and focus. When placed too early—or based solely on symmetry or aesthetics—it can easily interfere with finding the best speaker position.

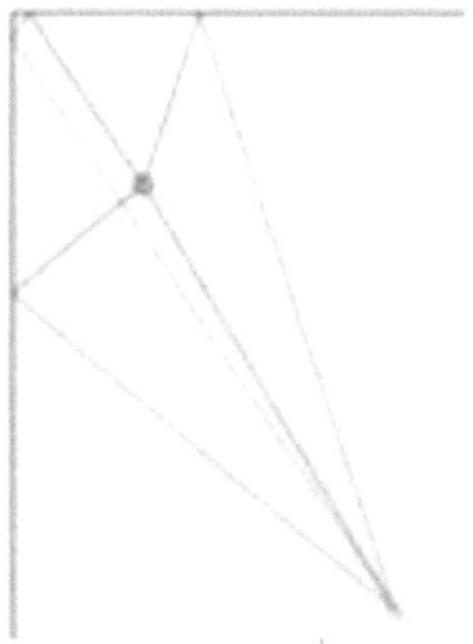

I want to emphasize once again that **properly optimized speaker placement in a normally furnished room does not require specialized acoustic treatment to sound exceptional**. If you choose to explore acoustic treatments, proceed deliberately and cautiously. It is very easy to overspend and, in some cases, to make the sound worse rather than better.

In Summary

1. **Sound is a wave** and obeys the principle of superposition—sound waves combine through addition and subtraction.
2. **All rooms have modes**, determined by the room's dimensions and shape.
3. **Speakers radiate sound in all directions**, and a portion of that sound is reflected back toward the listener.
4. **Out-of-phase reflections create frequency nulls** at specific frequencies.
5. **Sound waves can be absorbed or diffused** to shape frequency response and perceived clarity.
6. **All room modes exhibit high pressure in the four corners**, making corners the most universally appropriate locations for acoustic treatment.
7. **First reflection points** are the next most useful locations to evaluate for acoustic treatment—*after* speaker placement has been optimized.

The Most Important Part of a Room

In the previous section, we discussed dedicated rooms and acoustic treatment. One critical element of the room that is almost always overlooked is **the floor**. This is not a discussion about treating the floor acoustically, but about the floor itself. If the floor is unstable, the system will be unstable. We do not want speakers settling into the floor, nor do we want the floor moving beneath them. Here, we'll examine how several common flooring types—**solid wood, carpet, and tile**—affect speaker setup and long-term stability.

Solid Wood Floors

We tend to think of solid wood floors as rigid and immovable, but in reality, they move significantly with changes in **relative humidity (RH)**. A 40% change in room humidity can result in approximately a **6% change in the moisture content of the wood**. The surface finish does not act as a moisture barrier; it does not prevent the wood from absorbing and releasing moisture as humidity changes.

A 6% change in moisture content can result in roughly a **1/8-inch dimensional change in a 4-inch-wide board**. If all boards expanded and contracted uniformly—and if the speakers were resting perfectly on spike cups—the speakers might float harmlessly on top of this movement. In practice, however, wood floors do not move uniformly. Boards expand at different rates, and stresses accumulate unevenly.

In the context of speaker setup—where adjustments are often made in increments of **0.1 mm or less**—this amount of movement is enormous.

A Real-World Example

I once set up a very high-performance system in a dedicated room with a solid oak hardwood floor. The floor was nailed to plywood over a concrete slab, a common construction method. The system was optimized to an exceptionally high level in **January**, when humidity was very low. The sound was extraordinary.

By late **April**, the system sounded unrecognizable. Dynamics were severely suppressed. The soundstage had collapsed, with a large hole in the center. The presentation was flat and lifeless. The only thing that had changed was the season. As spring arrived, humidity increased, the wood floor expanded, and the speakers were slowly pushed out of their original positions. The setup itself had not "failed"—the floor had moved.

Tile Floors

Tile floors are common in some regions and offer one clear advantage: **dimensional stability**. Tile does not expand and contract with humidity in the way wood does. Notably, tile introduces its own set of problems for speaker setup—chief among them **unevenness** and **grout lines**.

Natural stone and even ceramic tile floors are rarely perfectly flat. Small height variations make precise, controlled movements difficult, and speakers often fail to sit evenly on all feet or spikes. In addition, grout lines are unavoidable. Sooner or later, a speaker will need to straddle a grout line, at which point it can become unstable or "hung up" in position. Inevitably—thanks to Murphy's law—the ideal speaker location ends up being right on top of one of these lines.

Smooth stone surfaces such as marble introduce another risk: **slippage**. Speakers can slide far too easily, especially when playing loud, dynamic material. Large speakers with powerful bass drivers can actually be pushed backward during heavy bass passages, which is completely unacceptable for maintaining alignment.

If speakers must be placed on a tile floor, choose tiles that are as **flat and consistent** as possible. Use **narrow grout lines**, and ensure the grout is nearly flush with the tile surface. Be especially cautious with polished or slippery finishes.

This Floor Will Never Work for Precision Speaker Placement

The Takeaway

Speaker placement precision depends on **mechanical stability**. Floors that move, compress, flex, or prevent consistent contact will undermine even the most careful setup. Before worrying about room treatments, diffusers, or absorbers, make sure the speakers are supported by a surface that allows them to **stay exactly where you put them**—not just today, but months from now.

Carpeted Floors and Speaker Stability

Unfortunately, the most common floor covering is **carpet**, and it presents serious challenges for precise speaker setup. There are essentially two ways people attempt to place speakers on carpet. The first—and most common—is placing the speakers directly on top of the carpet using feet or spikes that **do not penetrate** the carpet and padding. There is nothing beneficial about this approach.

There are two fundamental problems. First, both the carpet and the padding are **compliant materials**. Neither provides a stable reference surface while the speaker is playing music. In a stereo system, the left and right speakers are reproducing different signals, which means they are physically reacting differently. When the speakers are sitting on a compliant surface, they move **out of phase with the music**, smearing timing cues and damaging delicate high-frequency information.

Second, the speaker will **sink unevenly** into the carpet and padding over time. This makes the setup inherently unstable. Any careful positioning you achieve will slowly drift.

A Real-World Example

This problem is easy to demonstrate at audio shows. Having set up speakers at many of them, I can say with confidence that placing speakers directly on carpet—regardless of whether the carpet is thick or thin—results in **rapid sonic drift**. Thinner carpet settles faster; thicker carpet takes longer, but the outcome is the same.

At one recent show, the carpet was particularly thick. I completed the left speaker setup and began working on the right speaker when everyone decided to break for dinner. When I returned a few hours later, the sound was shockingly bad. I initially suspected the right speaker, but when I checked the left speaker, its **rake and azimuth had shifted significantly**. After correcting them, the right speaker setup proceeded normally.

The following morning, I checked both speakers again—and once more, their height, rake, and azimuth had drifted. This continued daily. By the third day the movement had slowed, but it was still clearly measurable and audible.

The takeaway is simple: unless you want to constantly re-adjust your speakers, compliant feet or **carpet cannot be the final support surface**.

The most stable surface is a **concrete slab**. In homes with wall-to-wall carpet over a slab foundation, the correct approach is to **spike through the carpet,** so the speaker feet make direct contact with the concrete.

If carpet is not required, **engineered wood flooring** is an excellent alternative. It is dimensionally stable, hard, and flat. **Bamboo flooring**, in particular, performs exceptionally well. It is stable, very flat, hard, and effective at dissipating vibrational energy.

Tile installed over a slab is also dimensionally stable, but grout lines can severely complicate setup. As discussed earlier, grout joints often prevent fine positioning and can make speakers unstable. If tile is used, grout lines should be narrow and nearly flush.

For homes built on **pier-and-beam foundations**, or second floors, the floor structure itself becomes the limiting factor. In these cases, the floor should be reinforced so it does not flex or bounce when walking near the speakers. Engineered wood flooring should be used instead of solid wood.

A Recommendation for Dedicated Rooms

If you are building or remodeling a dedicated listening room, there is one additional step worth considering. Before installing the final flooring, have the floor **floated**. This process levels the surface extremely precisely. The result is that the left and right speakers will naturally sit at identical heights, allowing nearly identical rake angles, improving dynamics and coherence without additional setup effort. Advice for Building a Dedicated Room

1. Do not build the room on a **second floor**, as this behaves like a pier-and-beam structure.
2. If pier-and-beam or a second floor is unavoidable, **reinforce the floor**. Adding a layer of cement board to the subfloor is a good approach.
3. Use **engineered wood flooring**, with bamboo being an excellent choice.
4. Avoid **compliant speaker feet**, such as rubber.
5. Do not allow speakers to "float" on carpet.

Speaker Feet: What Actually Works

Since flooring and speaker support are inseparable topics, it's worth addressing **speaker feet** directly. Speaker feet must meet two requirements:

1. **They must be adjustable**
2. **They must be stable**

Threaded feet are essential. Precise control of height, rake, and azimuth is impossible without them. Stability is equally critical. Feet made from rubber or other soft materials will move over time and do not provide a solid launch surface. Compliant materials blur bass response and reduce articulation.

Some speakers—and aftermarket products—use feet with **ball bearings**. While these can make positioning easier, the bearings can become displaced during repeated adjustments, causing noise. If your speakers use this type of foot, periodically lift the speaker slightly so the bearings can reseat before continuing adjustments.

Many people report improved bass when switching to compliant feet. Be careful with this conclusion. Most of these feet **raise the speaker**, and as discussed earlier, increased height alone improves bass articulation. To make a fair comparison between feet, always ensure the speaker is at **the same height** for both configurations.

Alternatives to Traditional Room Acoustic Treatment

Everything in a room influences the sound. Where objects are placed, what they are made of, and how they interact with sound waves all matter. Some objects—such as a fabric-covered sofa—**absorb** sound. Others—like glass windows or flat-screen televisions—**reflect** sound. Irregular objects, such as sculptures, can **scatter** sound. Still others can **resonate**, including windows, bowls, pottery, and even leather furniture.

We typically position speakers around the room's contents without giving much thought to those contents themselves. Yet any change within the room can and will affect the perceived sound to some degree. The higher the system's resolution, the more audible these changes become.

For most people, domestic harmony limits the extent to which a room can be filled with conventional acoustic panels. So what options remain? Interestingly, there are **non-traditional objects**—usually small—that can influence how sound is perceived in a room.

Some commercially available devices affect **perception rather than measurable acoustic response**. If measurements are taken with and without these devices, little or nothing changes in the data. It is common to hear the argument that if something cannot be measured, it must not be doing anything. I disagree with that assumption. The fact that a phenomenon is not easily measurable does not mean it is not perceptible. The way the human brain interprets sound—especially spatial cues and subtle timing information—is still not fully understood, and this remains an active area of research.

The devices I am referring to are **resonators**. Several companies manufacture versions of these devices, including **Acoustic System International**, **Neodio**, **Synergistic Research**, and **HighEnd Novum**, among others. These resonators come in a variety of sizes, shapes, and materials, and each interacts differently with the sound field.

Their effects are not universal and cannot be predicted in advance. They must be **experimented with**, much like speaker placement itself. Different resonators can emphasize different qualities—more body, greater focus, increased air, or improved spatial definition. Placement is critical. Because these devices operate by resonance, their effectiveness depends strongly on where they are positioned in the room. Certain locations will excite their behavior far more than others.

I raise this topic because, in my experience, these devices can **meaningfully alter the perceived soundstage and coherence of a system** once everything else has been optimized. I encourage you to explore them only after speaker placement and system setup are complete. They are not foundational tools—but they can sometimes be the final refinement that resolves the last lingering issue in how music presents itself in your room.

Used thoughtfully, they may surprise you.

Summary: Understanding the Listening Room

In Chapter 11, we broaden the discussion from loudspeaker placement to the larger acoustic environment in which the system operates. While the Precision Setup Method has emphasized careful positioning of the speakers and listening seat, the room itself remains an unavoidable part of the playback system. Every surface in the listening space reflects, absorbs, and shapes the sound that ultimately reaches our ears.

This chapter introduces the essential principles of room acoustics—how sound waves interact with walls, floors, ceilings, and furnishings, and how these interactions influence tonal balance, spatial clarity, and perceived realism. Understanding these interactions allows the listener to make thoughtful decisions about the listening environment rather than relying on guesswork or excessive treatment.

Throughout this guide, one principle has remained consistent: exceptional sound begins with correct speaker placement, not aggressive acoustic treatment. When speakers are properly aligned and positioned relative to the room and listening seat, many acoustic problems diminish naturally. In contrast, when treatment is applied before the system itself is optimized, it often masks symptoms rather than solving the underlying cause.

For this reason, room treatments should be viewed as refinement tools rather than primary solutions. Carefully applied acoustic treatments can polish an already well-aligned system by reducing unwanted reflections, managing bass energy, and improving clarity. When misapplied, however, they can dampen musical energy, reduce spatial openness, and diminish the natural character of the recording.

This chapter also introduces the References section that follows. The works cited there provide deeper scientific and technical exploration of the topics discussed throughout this book, including spatial hearing, psychoacoustics, room acoustics, and sound reproduction. These references are not required reading in order to apply the Precision Setup Method, but they offer valuable resources for readers who wish to explore the underlying research in greater depth.

Together, this chapter and the accompanying references place the practical setup techniques of this guide within a broader scientific context, linking real-world listening experience with the established research on how humans perceive sound and how acoustic spaces influence that perception.

The goal is not to overwhelm the listener with theory, but to provide enough understanding to make intelligent decisions about the listening environment while preserving the musical realism that careful speaker alignment can reveal.

THANK YOU!

Thank you for joining me on this journey. I wish you the very best and hope this guide has helped you finally understand how to position speakers, so they sound like music.

Happy listening, and good luck!

APPENDICES

System Changes and Component Accommodations

Once a system is fully optimized, **timing accuracy between the electronics, speakers, and room reaches a very high level**. At that point, any change—no matter how small—will affect either timing or the way the speakers interact with the room. This includes changes that are generally considered upgrades. A new component may improve certain aspects of the sound, yet listening to works such as *Beethoven's Piano Concerto No. 1* or *Shostakovich's Symphony No. 1* can suddenly reveal problems: the performance feels disjointed, violins lose delicacy, instruments no longer play together, and the sound takes on a ringy or dissonant character.

These issues arise because the new equipment or cabling has a **different timing characteristic** than what the system was originally optimized around. The speakers were precisely positioned to work with a specific set of electronics. When that relationship changes, the balance shifts.

Every component has its own timing behavior, described technically as a **transfer function**. When a component is replaced, the system inherits a new timing relationship. As a result, the speaker setup must be adjusted to compensate. When this adjustment is made correctly, you retain the original timing coherence while gaining the benefits of the new component. In fact, the result is often better than before, because a lower noise floor allows timing cues to be heard more clearly.

If this sounds abstract, the amplifier provides an easy example.

The loudspeaker is mechanically coupled to the room—its modes, boundaries, and reflections. Earlier in the process, we minimized those interactions by using *"Chant"* to establish correct placement. The amplifier directly drives the woofers that energize the room, and different amplifiers interact with those woofers in very different ways.

Amplifiers vary in **output impedance (damping factor)**, which determines how firmly they control woofer motion. They also differ in the **amount and type of negative feedback**, which further influences how the amplifier responds to the speaker load. If the speakers were originally set up using a tube amplifier and are then driven by a high-powered solid-state amplifier, the level of woofer control will change dramatically. As a result, the way the speakers excite the room will also change.

In this scenario, bass may suddenly sound overpowering, and when playing *"Chant"* you may find yourself asking, *"Where did the growl go?"* The growl has not disappeared—it is simply no longer aligned. If you repeat the setup process, the growl will reappear.

This effect occurs with **every system change**, to varying degrees. If this is still difficult to accept, try a simple experiment. Take a system that is fully dialed in and swap a single component. Then listen to Beethoven or Shostakovich. You will hear that the performance has lost its cohesion.

The encouraging news is that once a system is truly optimized, **the benefits of better equipment become very easy to hear**. Reductions in noise floor are immediately obvious. Improvements from shelves, footers, and other mechanical interfaces become much more apparent. The only downside is that to fully realize those gains, **the setup process must be revisited**.

That is not a flaw in the method—it is evidence that the system is resolving enough to reveal the changes.

Speakers With Adjustable Modules

Some speakers are designed and manufactured with one or more of the drivers independently adjustable. When setting up a pair of speakers with **movable driver modules**, it is essential to keep the settings on both speakers **matched at all times**. Even though the right speaker may be turned away during setup, every initial adjustment made to the left speaker must be duplicated on the corresponding elements of the right speaker. This consistency has a profound effect on the perceived soundstage and overall coherence. As you work through adjustments on the left speaker, any change to a module—no matter how small—should be immediately applied to the right speaker as well.

For most loudspeakers, the relationship between drivers is fixed. The woofer, midrange, and tweeter are housed in a single cabinet and move together as one unit. These speakers present **six degrees of freedom**, which we manipulate to achieve a coherent sound field. In contrast, some designs allow **independent adjustment of individual drivers**. In some cases, only the tweeter can be moved, typically forward or backward. Other designs permit multiple drivers to be adjusted independently, both front-to-back and laterally.

Each independently adjustable driver—and each direction of movement—adds additional **degrees of freedom** to the setup process. While this flexibility can offer greater potential precision, it also increases complexity. Every added degree of freedom becomes another variable that must be correct for the system to sound right. It is easy for this process to become overwhelming if it is not approached methodically.

If you have chosen speakers with movable driver modules, that choice is perfectly valid. Just be aware that these designs demand greater discipline during setup. Careful tracking of adjustments and strict symmetry between left and right speakers are essential to realizing their full potential. Typically, the manufacturer's manual will include a set of parameters to measure, along with predefined driver settings based on those measurements. In principle, you could simply follow the instructions, set each driver to the recommended position, and then treat the speaker as a single unit during setup. This approach will usually produce a **reasonable** result. However, if the goal is optimal performance, several important issues must be considered.

Measurement Accuracy

The first issue is **measurement precision**. Manuals often specify very small offsets. For example, a manufacturer may recommend that the midrange driver be positioned so its distance to your ear is **22.5 mm greater** than that of the tweeter. Accurately measuring such a small difference is difficult in practice. Even high-quality laser distance meters typically have a stated accuracy of **±1 mm**, and less expensive units are often worse. These tools can get you close, but they cannot deliver absolute precision. As a result, **final positioning must still be refined by ear**, regardless of how carefully the measurements are taken.

Coarse Adjustment Ranges

The second issue is that the recommended settings often span **very large ranges**. A manual might specify one setting if the listening distance is 1 foot and a completely different setting if the distance is 2 feet. That leaves an obvious question: what if the actual distance is **1 foot 6 inches**? The manual provides no meaningful guidance in that case. The only reliable way to determine the correct setting is to **listen to both options** and decide which produces the better result.

Unintended Degrees of Freedom

The third issue involves **unintended movement** of driver modules. A module may be designed to move in only one degree of freedom, yet in practice there is often enough mechanical play for movement in other directions. Consider a midrange module mounted on spikes that sit in detents. Even if the intended adjustment is vertical, the module may still rotate slightly, effectively introducing toe-in or toe-out. If the design assumes the module is pointing straight ahead, this unintended rotation must be controlled. In practice, ensuring precise alignment of individual modules is often difficult and sometimes frustrating.

Manufacturing Tolerances

The final major issue is **manufacturing tolerance**. It is not enough for one speaker to be correct—the second speaker must match it exactly. Even very small tolerances can create meaningful differences. Consider a tweeter mounted on a linear slide attached to a vertical rail with two screws. If the screw holes are offset by just **0.2 mm**, the rail may tilt slightly. If the left speaker is assembled so that this tilt angles the tweeter upward, and the right speaker is assembled with the rail flipped 180 degrees, the tweeter on that side will tilt downward by the same amount. These discrepancies must be identified and compensated for during setup.

When multiple drivers each have multiple degrees of freedom, these tolerances accumulate quickly, creating a surprisingly large number of potential integration errors.

The Practical Reality

Because real-world manufacturing is imperfect and measurement accuracy is limited, there is no alternative but to **integrate all drivers using both the intended and unintended degrees of freedom**. Describing a universal method for integrating modular speakers is beyond the scope of this book, as each design must be approached individually. However, the same **sonic cues** used to integrate a complete loudspeaker apply here as well—only now they are applied **driver by driver** until the speaker behaves as a coherent whole.

The Trouble with Sibilants and Sibilance

Have you ever wondered how the first word in Diana Krall's rendition of *'S Wonderful* is actually supposed to sound? That simple 'S' sound is a key to understanding both what is correct and how to identify—and refine—what might be wrong. It's one of the few recordings that offers a useful, verifiable listening reference point. Why? Because the song begins with a single sibilant consonant that transitions into a warm, pitched note full of harmonic presence. But how do you know if what your system is playing is correct? And even if you like the current sound, could it be improved in a way you don't currently perceive?

This is where conventional wisdom about "brightness" and "harshness" in playback often goes astray. We typically look for trouble in those pesky "esses," but we've been somewhat misguided in our description and approach. The problem starts with a fundamental misunderstanding: the words we use—sibilant and sibilance—describe related but distinct phenomena, yet we often treat them interchangeably. For the purpose music playback, I'm using sibilant in the standard phonetic sense for consonants such as S, Z, ʃ, 3, Tʃ, and D3. When discussing playback problems and cumulative high-frequency effects, I'll refer to excess sibilant energy, high-frequency smear, or—more broadly—the accumulated or exaggerated high-frequency character that emerges from multiple sources. This distinction matters profoundly for how we reproduce and listen to music.

What Are Sibilants?

Without diving into exhaustive linguistic detail, sibilants are naturally occurring consonant sounds that help form the words we use in everyday speech. Linguistically, sibilants appear in two forms: as fricatives (the "s" and "z" sounds) and as components of affricates (the "ch" and "j" sounds). These are produced by directing a stream of air through a narrow constriction—typically between the tongue and teeth, or between the lips—creating turbulence and noise.

Sibilants occupy roughly the 2–8 kHz band, though the exact range varies substantially by consonant, talker, language, and recording chain. This range covers a significant portion of where our hearing is particularly sensitive. Because we are so attuned to this frequency region, sibilants help shape articulation, timing, and intent in speech. Different languages, cultures, and individuals use sibilants in a multitude of ways.

Sibilants are a subset of what linguists call stridents, but they're distinguished by their intensity and higher pitch relative to other fricative sounds. Here's where precision matters: many sibilants, especially voiceless ones such as S, are dominated by aperiodic frication noise with no fundamental frequency. But voiced sibilants such as Z and 3 combine that frication with simultaneous vocal cord vibration. In other words, the vocal cords can vibrate while turbulent sibilant noise is present. So it is not correct to say that ALL sibilants are pure aperiodic noise, but some are. This distinction is important because it means sibilants exist along a spectrum from purely noise like to semi-voiced, depending on the consonant itself.

The Physiology of Attention: The "Psst" Example

Consider a simple, deliberate example: the "psst" sound when trying to get someone's attention. Here, you deliberately force air through your teeth to create a hissing sound. We are sensitive to the frequency band that produces the sss sound, so it's easy for us to hear. But notice what happens before it: there's a sudden burst of air as your lips release, creating a transient acoustic event that's spectrally distinct from the sibilant hiss that follows. That transition from the plosive (the lip release) to the fricative (the hissing), though seemingly insignificant, is physiologically meaningful to how we perceive the "psst" sound. In the P of "psst" your lips briefly block the airflow and then releases it, creating a transient burst before the frication begins. This release, what researchers call perceptual salience, grabs our attention much better than just the sss sound alone. Without the sharp plosive that precedes the sss, the sound lacks a clearly defined onset; it's vague and without motivation. The addition of the P to the sss provides a sharp, transient element that our brain decodes as: pay attention! It also provides a solid temporal beginning, which in music, is critical.

You might wonder: how does this relate to speaker setup? Here's the connection. Research in psychoacoustics shows that temporal precision in transient onsets, the sharpness of that explosive burst, affects how readily a sound captures one's attention and integrates into memory. These temporal relationships directly affect how we process and segment the sounds that we recognize as music. It's important to at least be aware of how these seemingly insignificant details in timing and character have a meaningful impact on what our brains are actually doing with the sounds we hear.

Sibilants in Musical Context

In music, the impact of sibilants scales with the number of sources and the velocity of air movement. When 100 singers produce a note beginning with the sibilant "s," the cumulative air velocity creates a pronounced effect. But notice something crucial: when those same singers produce that sibilant at p or pp (soft dynamics), the sibilant is still present but quieter—not harsh. At f or ff dynamics, the same sibilants intensify proportionally. This is not a flaw; it's physics. The acoustic energy scales directly with airflow velocity. Sibilants are also routinely used as deliberate artistic devices. A composer might write a passage where sibilants add tension, urgency, or clarity to text. A vocalist might emphasize sibilants to create presence or intimacy. These are intentional choices, not accidents.

The key point: sibilants are natural, discrete acoustic events. They begin and end. They have a defined envelope. They are not something to eliminate from music—they are something to preserve and present faithfully.

What Is Sibilance?

Sibilance, as I'm using the term, refers to the cumulative or exaggerated high-frequency character that emerges from the interaction of multiple sources of acoustic energy. Unlike the discrete nature of a single sibilant consonant, this cumulative effect can arise compositionally, performatively, or as a byproduct of the recording and playback chain.

Sibilance in Language and Literature

In literature, sibilance operates as a deliberate rhetorical device: a writer strings together sibilant consonants – "Sally sells seashells by the seashore" – using the natural perceptual salience of these sounds to create rhythm and mood. But the experience as a listener is the acoustic result: the cumulative effect of those repeated sibilants, each one reinforcing the others, creates a unified hissing quality that shapes pacing and atmosphere.

Sibilance in Music: Three Sources

In music, this cumulative or exaggerated high-frequency character can arise in several ways.

Compositionally: In classical choral orchestration, sibilants can be deployed with great precision. The sopranos and upper strings naturally occupy a similar frequency space. A composer can deliberately reinforce or suppress the interaction between these sections to build or release tension. The same principle applies within the orchestra itself—different instrumental sections can create cumulative high-frequency effects through their interaction, and this becomes a tool for emotional or dramatic effect.

Performatively: A jazz drummer using wire brushes to lay down a continuous "shhhhh" texture behind a vocalist creates a broadband sibilant environment. This contributes to mood, energy, and context. The interaction between the vocalist's deliberate sibilant technique and the drummer's brush pattern creates something neither produces alone—a unified sonic statement. Consider John Coltrane's approach: his tenor saxophone naturally produces a complex harmonic spectrum. His mouthpiece and embouchure techniques deliberately introduce breathy, textured characteristics—including sibilant-like energy—into his instrumental sound, blending pitched content with high-frequency noise. The result is a distinctive sonic signature. This wasn't incidental; it was integral to his musical voice.

As a system artifact: High-frequency exaggeration or harshness can also emerge as an unintended consequence of the recording, mastering, and playback chain. This is the sibilance problem most audio discussions focus on—the character that can cause fatigue or distraction. Understanding this source is crucial for listeners and engineers alike.

Why This Distinction Matters

The reason we must differentiate between sibilants (the consonants themselves) and the sibilance problem (cumulative exaggeration across sources) is not semantic; it's practical and perceptual. When reproducing recorded music, we need to distinguish between:

1. Sibilants that are actually on the recording — the artist's intentional use of these consonants.
2. High-frequency exaggeration created by the recording process — artifacts introduced during capture or mastering.
3. High-frequency exaggeration created by our playback system — colorations introduced by speakers, room acoustics, or component interactions.

Conflating these three sources leads to poor decisions about how to address perceived "brightness" or "harshness" in playback. Understanding these distinctions is the foundation for listening critically to your system.

The Recording Chain: Capturing Sibilants Faithfully

Recording engineers face real constraints when capturing vocals with sibilants. Strong sibilants can be exaggerated by microphone choice, placement, capsule voicing, and downstream compression. In extreme cases, spectral over-emphasis can occur—though modern vocal condensers often handle enormous SPL without literal diaphragm clipping. (Neuman's KMS 104, for example, is rated to 150 dB SPL.) More commonly, the problem is not overload itself, but spectral over-emphasis created by the microphone's proximity effect, the vocalist's technique, or aggressive compression or saturation in the signal chain.

Professional recording engineers address sibilant challenges through multiple techniques: microphone selection, proximity (distance from source), angle, acoustically transparent shields, and—increasingly—electronic de-essers applied in real time or post-production. These techniques vary in their effectiveness and in their sonic consequences. Some preserve the natural character of the sibilant; others smooth or suppress it in ways that can rob the performance of clarity and presence.

The reality: many commercial recordings involve some tradeoff in sibilance handling, but the extent varies widely with the singer, microphone, arrangement, and production choices. Good recording engineers understand the cost of aggressive de-essing: a sibilant that's been electronically tamed often loses the transient clarity that makes it feel real and present. The best recordings minimize this damage through technique rather than fix it through processing.

For the listener, this means recognizing that some degree of sibilant handling in commercial recordings is common. The question becomes: is what we're hearing a faithful reproduction of what the recording engineer captured, or is our playback system adding further distortion?

The Audio System Challenge: Distinguishing Sibilants from Sibilance

This is where your audio system's setup becomes crucial. Because sibilants and high-frequency effects are predominantly upper-midrange and treble phenomena, addressing them involves primarily the tweeters and upper-midrange drivers. But here's the critical distinction: a sibilant should sound like what it is—a discrete, natural consonant. Deliberate artistic sibilance should help reveal the interaction and intention behind the music.

A Concrete Example: The Hi-Hat Cymbal

Consider the high-hat cymbal. Is the hi-hat bright? Yes—by design. A jazz drummer's hi-hat is supposed to be bright and present. But there's a difference between bright clarity and harsh smearing.

Ask yourself: Am I hearing the natural brightness of the instrument, or am I hearing an artifact—excessive high-frequency character created by tweeter interaction, or phase misalignment between tweeters and midrange drivers? These sound superficially similar but have different causes and solutions.

If what you hear is harsh rather than bright, possible causes include poor left-right matching between channels, phase misalignment between drivers, early reflections bouncing into the listening position, off-axis response problems, or the room's interaction with high-frequency energy. One possible playback-related cause is inadequate coherence between tweeters. But similar symptoms can also arise from the recording itself, from microphone technique, from early reflections in your listening space, or from room-response problems.

The solution isn't to blindly suppress the top end or add absorption. Once you've identified the actual cause, valid remedies can include adjusting speaker/listener geometry, improving channel matching, refining crossover integration, treating early reflections, addressing room-response problems, or—in specific cases—modest high-frequency adjustment. The principle is diagnosis first, treatment second.

When you get the system dialed in correctly, what previously sounded harsh is often revealed to be beautiful detail. The hi-hat is still bright—that's what it's supposed to be—but now it's also clear and articulate.

The Sibilant: A Case Study

Now let's examine a sibilant that can reveal system performance—one where the system is failing to preserve what the artist actually did.

Diana Krall's opening word in 'S Wonderful begins with a voiceless sibilant "s" that transitions into the voiced /w/ and then into the sung vowel. On a well-behaved system, that handoff from frication-dominant energy to the voiced, harmonic material that follows can be a useful playback cue. I offer it as a practical listening reference, not as a universal physiological rule.

Here's why it works as a diagnostic: This particular recording offers a clean, verifiable moment where you can listen for the system's ability to preserve transient detail and clarity. If your system fails to preserve this transition—if the sibilant is smeared or drawn out, or if it doesn't have clear definition before the /w/ begins—something important is lost. You no longer hear the artist's technique. You no longer hear the deliberate control and artistry of the performance.

The sibilant bleeds into the sustained material, diminishing the performance's impact.

What causes this degradation? One possible cause is misalignment between the two tweeters. If the left tweeter's high-frequency response peaks at a different frequency or arrives at a slightly different time than the right tweeter, the coherence collapses. Each speaker is producing a slightly different version of that initial sibilant, and they're not synchronizing cleanly. The result is a smeared, extended version of what should be a quick, clean acoustic event.

Another possibility: early reflections are disrupting the transient presentation, or off-axis response is coloring the high-frequency content in a way that obscures detail. The fix depends on identifying the actual cause.

In many cases—particularly in well-treated rooms with properly aligned speakers—optimizing tweeter coherence and phase integration can reveal vastly more detail throughout the music. A covered-up transition, a collapsed dynamic, a smeared sibilant—these areas are often where the music actually lives. Clean them up, and you'll be surprised at what else is revealed.

Listening Forward

The distinction between sibilants and the cumulative high-frequency character that can emerge from multiple sources is not merely technical. It's about how we listen. When you hear what sounds like" brightness" or "harshness," ask yourself:

Is this the natural quality of the source material, or is it an artifact of my system? Is this a sibilant that's supposed to be there, or is it excessive high-frequency character created by my speakers and room?

The answer changes how you address the problem—and whether you should address it at all.

A properly set-up audio system doesn't apologize for high-frequency content. It reproduces it faithfully. Sibilants are clear and discrete. Intentional cumulative high-frequency effects serve the music. And brightness that's supposed to be there remains bright, articulate, and beautiful. The power in understanding system- or program-material-induced high-frequency problems is the choice to act or not to act. But now you have a clearer idea of what you're actually acting upon. That's what we're listening for.

Further Thoughts on Timing and Speakers

Speakers come in all shapes and sizes. What they all have in common is some type of membrane or medium that vibrates in response to an input signal. In an ideal situation, the membrane would instantly move in response to the signal and stop immediately when the signal is no longer present. In the real world, this doesn't happen. There are several reasons for this, and all contribute to the time smearing of the signal. We will explore some of the reasons for this.

A driver may have undamped resonances. This will cause the driver to vibrate at specific frequencies long after the input signal stops. Gross timing errors will result as the driver does not leave enough space before the next note. Any note played at or near one of these resonant frequencies will sound "bloomy" or "ringy". For example, that piano note will sound blurry and unclear when the other notes around it sound perfectly clear and distinct.

The driver material may be too massive or heavy for the driving motor. This will mean the driver is slow to start and slow to stop. In the worst case, this could make any fast-paced music sound lethargic. If someone listens almost exclusively to vocals, this may not sound bad. It can give the sound a warm glow. But getting piano or other music to sound "right" will be challenging.

There could be issues with the crossover design. One of the biggest problems is poor parts quality. Many engineers believe that all capacitors with a given capacitance are the same. Likewise, for inductors and resistors. Inductors and capacitors store and release electrical energy. It seems some types of these components release this energy slower than others. Another issue with crossover design is phase shifts occurring at the crossover point. Another word for phase shift is time delay. This means the crossover is building in a time delay between two drivers. Another issue with crossover design is part tolerances. If the manufacturer is not careful, the two speakers that make up the stereo pair can have very different values for the capacitors, inductors and resistors that make up the crossover. This will result in poor acoustic matching between the speaker and effect stereo imaging negatively.

Poor cabinet design can result in vibration of the cabinet walls. This vibration is similar to driver ringing. These resonant frequencies continue to play after the note has stopped. These vibrations are clearly audible and cause the bass to sound muddy or wooly if they are strong enough. Even if they are not audible, they affect the response of the driver. Some speaker designs view this as such a problem that they do away with the closed cabinet altogether. Examples of this are open baffle speakers, panel speakers and horn speakers.

Resolving all of the mentioned issues and more is what paying more money for a speaker should be about. This is not always the case, but it is generally true. We can still optimize the speakers we have but the optimization process can only be pushed as far as the timing accuracy of the speaker. When picking a speaker, there are a lot of choices on the market. I encourage anyone to do some research before spending their hard-earned money. Investigate how the speaker is built and how much engineering effort has gone into its design. Ask questions like:

1. What material is used to construct the cabinet? How thick is it? How much internal bracing? Thick-walled, highly braced boxes are usually desirable. The weight of a speaker can be a good indication of solid construction.
2. What material are the drivers made from? Are all of the drivers made from the same material? Metal drivers have a reputation for ringing. Paper material is self-damped and typically has little ringing. I am not saying paper is better than aluminum, but care needs to be taken in how the driver is designed and used.
3. Are good-quality crossover parts used? Electrolytic capacitors and iron core inductors for the mid-range driver are not good choices.
4. Many of the design choices are inside the speaker and not seen. I encourage you to look past the shiny, high gloss exterior and ensure what you are buying is well engineered.

Tube Group Delay

This discussion is not intended to argue the merits of solid-state versus tube electronics. Rather, it is meant to draw attention to a significant pitfall to which owners and admirers of tube-based components are particularly susceptible: how well the group delay characteristics of the tubes are matched between channels.

Group delay was discussed in detail in Chapter 10 in the context of subwoofer integration, but the concept applies equally to electronic components. Any time a frequency-dependent phase shift occurs, group delay is present. Throughout this section, the term phase shift is used to indicate that not all frequencies pass through a stereo system at the same speed.

When we speak of a "system," it is important to recognize that each component within that system has its own unique group delay characteristics. Within the audio band, the group delay behavior of an amplifier can be thought of as a snapshot of how—and when—each frequency moves from input to output. As components are chained together, increasing system complexity, their individual group delay characteristics accumulate and interact, altering the overall temporal behavior of the system. In an ideal world, group delay would be zero. At present, that ideal remains unattainable.

In general, the more linear—or "clean"—an amplification system is across a given frequency band, the less phase shift it introduces. Well-designed, wide-bandwidth solid-state amplifiers (greater than roughly 250 kHz) tend to maintain relatively linear phase behavior throughout the audio band. Even so, they are not perfect, which helps explain why amplifiers can sound and behave differently from one another. One advantage of solid-state devices is long-term stability: transistors tend to preserve their phase behavior consistently over many years.

This is not to suggest that tube amplifiers cannot exhibit excellent phase behavior in the audio band. They can. Granted, vacuum tubes are inherently less uniform than solid-state devices, making channel-to-channel matching more difficult. Compounding this issue is the fact that tubes change as they age, gradually drifting away from their intended operating characteristics. This brings us to the subject of tube matching.

Consider an amplifier in which each channel relies on a tube performing the same function. If one tube behaves slightly differently from its counterpart within a particular frequency range, even by a small amount, the result is asymmetric phase shift between channels in that band.

This is not intended to be a deep technical discussion of the mechanisms that cause phase shifts. What matters here is the consequence. A phase shift alters the output impedance of a tube, which in turn changes how the following stage is driven. That change affects the operating point of the next stage, which then alters its own phase behavior. The cumulative result is, at minimum, a mild temporal asymmetry between channels and, in the worst case, a group delay curve that causes a particular frequency band to arrive 20 microseconds after the bulk of the musical content.

In real-world terms, a 20-microsecond delay corresponds to approximately 7 millimeters of acoustic distance. When attempting to precisely align loudspeakers, a 7 mm discrepancy—especially if it occurs in the critical 1 kHz to 3 kHz region—cannot be corrected. This is because the error exists simultaneously in both the time and amplitude domains.

When selecting replacement tubes or experimenting with alternatives, the following practices are strongly recommended:

- Purchase a modern tube tester and periodically measure the condition of your tubes.
- When removing tubes, clearly mark them so they can be returned to the same component and the same socket position.
- Always purchase and install tubes in matched pairs.
- Acquire multiple matched pairs and experiment with different pairings to determine which combination performs best in your system.
- If a circuit uses two tubes for the same function, always try swapping them. In my experience, it is rare for this swap not to reveal one configuration that sounds clearly better than the other.

Music Notes and the Frequency Range of Instruments

Instrument	Low Note	Low Frequency	High Note	High Frequency
Piano	A0	27	C8	4186
Double Bass	E1	41	B3	247
Bass Guitar	E1	41	E5	659
Cello	C2	65	A6	1760
Guitar	E2	82	E6	1319
Clarinet	D3	147	B6	1865
Trumpet	E3	165	B5	988
Violin	G3	196	E7	2637
Flute	C4	262	C7	2093
Vocal				
Bass	E2	82	F4	349
Baritone	F2	89	G4	392
Tenor	C3	131	B4	494
Alto	F3	175	F5	699
Soprano	B3	247	D6	1175

Adding Sound Levels

If two independent sound sources are both playing, we may need to know the resulting loudness. This consideration is mostly relevant for subwoofers. If two speakers are playing at the same loudness, the resulting sound will be 3dB higher than if only one were playing. This implies that if we set up two subwoofers, we should aim to have each 3dB down from our intended target. When both subwoofers are turned on, the resulting sound will be 3dB louder.

Below is a table that shows typical differences and the resulting loudness level.

0 dB difference, then the result is 3 dB higher

5 dB difference, then the result is 1.2 dB higher

10 dB difference, then the result is 0.4 dB higher

15 dB difference, then the result is 0.15 dB higher

20 dB difference, then the result is 0.04 dB higher

TODD'S ACKNOWLEDGMENTS

This book would not have been possible without the input and encouragement of the people who helped me find my way in audio. I would like to thank each of you.

To my awesome wife, Sally, thank you for your constant love, support and grounding. You were there that cool October morning on our adventure to find our first pair of real speakers. You have been there every step of the way. Without you this book would not exist. Thank you for indulging my passion for this hobby.

To my uncle Mark, you always had a cool stereo while I was growing up. I fondly remember listening to many records on it. It was you who bought me that Sony Walkman and a copy of Michael Jackson's "Thriller" for Christmas. And you would make mix tapes for me to listen to while I played video games. These early gestures were the spark that set off the conflagration. Thank you for everything you have done throughout my life.

To Chuck, my first audio dealer, thank you for being so patient and kind. During each of my many visits to your store you always greeted me and answered my questions with enthusiasm and passion. I feel you genuinely tried to help me find my way. Thank you for pulling back the curtain and showing me the first glimpses at what is possible with two-channel audio.

To Paul, the first truly dedicated audiophile I encountered, thank you for showing me what is possible with a modest budget and a system where every detail is considered. Thank you for introducing me to horn speakers as something that could produce remarkable detail and believability. I will always remember listening to your system and our conversations about audio. Those experiences stay with me.

To Stirling, the master of detail, thank you for your willingness to teach me. I am certain it was frustrating. I realize it is difficult to explain to someone what they should hear, but you were always willing to put in a few more repetitions until I finally got it. When I think back on all the time we spent talking about and working through the details of maximizing sound reproduction, I am in awe of the amount of information you hold in your mind. I am grateful that you were willing to share so much of it.

To this book's early readers, Bo, Al, Adam and Jonathan, thank you for taking time out of your busy schedules to provide valuable feedback. Your thoughtful consideration led to increased clarity and purpose. I believe your suggestions will make it easier for future readers to comprehend the strange concepts presented in these pages. I wish each of you the very best in your continued journey in music enjoyment through high-fidelity audio.

STIRLING'S ACKNOWLEDGEMENTS

Many people shaped the journey that led to my part of this book. I'm grateful to all of them.

To my father, who played jazz on his Fisher stereo on Sunday afternoons — loudly enough that I complained to my mother to make him turn it down. Whatever he was doing stuck with me. And to my mother, Alice B. Trayle — one of the first women to graduate from UC Berkeley's optometry school, Class of 1944, admitted at sixteen — who moved our family to Petaluma in 1964 to open her practice, and who showed me early what it looks like to build something on the strength of your own expertise.

To my daughter Veronica, who has held her vision without compromise since childhood, and who has not yet found the key that unlocks it. I recognize that persistence — and I believe you will find it. I didn't find mine until I was 55.

To Roy Gregory — who saw what I was capable of before I had committed to making it a business, told me plainly that the industry needed it, and then made the case publicly with a review unlike any he had written before. The debt is foundational.

To Patrice Ryan and Julie Terry, my earliest clients who became some of my closest friends — who were the first to show me something I hadn't anticipated: that the greatest reward in this work wasn't technical, it was witnessing what music could suddenly do for someone whose system had finally been set free. I felt it first with them. I've felt it with every client since.

To Todd Binnix — computational physicist, music lover, and long-time audiophile — who did two things this book required and that I could not have done alone: he analyzed a methodology I had built through decades of practice and gave it language precise enough to teach, and he conducted and wrote the scientific research that underpins it. What he gave this project goes well beyond collaboration.

And to my partner Carol Messer — who, when Audio Systems Optimized was slow to find its footing, gave me six months to prove it could work before she would ask me to consider another path. That deadline motivated me beyond anything my own ambition could have managed. There were points I would have walked away entirely. That I didn't is largely her doing.

QR CODE TO AUTHORS' WEBSITE

The website linked to the QR code below is provided for future reference material. It will contain errata, subject clarifications, related articles pertaining to speaker setup and guiding videos.

REFERENCES

The ideas presented throughout this book draw from both practical listening experience and a substantial body of scientific research into acoustics, psychoacoustics, and human hearing. The references that follow provide entry points into that larger body of knowledge. They are included for readers who wish to explore the underlying science more deeply or examine the studies and texts that have shaped our understanding of spatial hearing, sound reproduction, and acoustic environments.

While the Precision Setup Method is designed to be learned primarily through careful listening and hands-on experimentation, these works offer valuable context and perspective for those interested in the broader theoretical foundations of the subject.

- Kunchur, M. N. "The Human Auditory System and Audio." Applied Acoustics (2023).
- Opstal, A. John van. The Auditory System and Human Sound Localization. Amsterdam: Academic Press, 2016.
- Plack, Christopher J. The Sense of Hearing, 4th ed. London: Routledge. ISBN 978-1-032-29951-8.
- Lyon, Richard F. Human and Machine Hearing: Extracting Meaning from Sound. Cambridge: Cambridge University Press. ISBN 978-1-107-00753-6.
- Strutt, John William (Lord Rayleigh). The Theory of Sound. Cambridge: Cambridge University Press. ISBN 978-1139058087.
- Kunchur, Milind N. Electronic Journal Technical Acoustics, "Audibility of Temporal Smearing and Time Misalignment of Acoustic Signals." (August 2007).
- Thavam, T., and M. Dietz. "Smallest Perceivable Interaural Time Differences." Journal of the Acoustical Society of America (2019).
- Blauert, Jens. Spatial Hearing: The Psychophysics of Human Sound Localization. Cambridge, MA: MIT Press. ISBN 978-0262024136.

- De Sena, Enzo, et al. "Localization Uncertainty in Time-Amplitude Stereophonic Reproduction." IEEE/ACM Transactions on Audio, Speech, and Language Processing (2019).
- Koch, Christian. "Hearing Beyond the Limit: Measurement, Perception and Impact of Infrasound and Ultrasonic Noise." Paper presented at the 12th ICBEN Congress on Noise as a Public Health Problem, 2017.
- Litovsky, Ruth Y., H. Steven Colburn, William A. Yost, and Sandra J. Guzman. "The Precedence Effect." Journal of the Acoustical Society of America 106, no. 4 (1999): 1633–1654.
- Jeffress, Lloyd A. "A Place Theory of Sound Localization." Journal of Comparative and Physiological Psychology 41, no. 1 (1948): 35.
- Gaik, Wolfgang. "Combined Evaluation of Interaural Time and Intensity Differences: Psychoacoustic Results and Computer Modeling." Journal of the Acoustical Society of America 94, no. 1 (1993): 98–110.
- Pulkki, Ville, and Timo Hirvonen. "Localization of Virtual Sources in Multichannel Audio Reproduction." IEEE Transactions on Audio, Speech, and Language Processing 13, no. 1 (2005): 105–119.
- Faller, Christof, and Juha Merimaa. "Source Localization in Complex Listening Situations: Selection of Binaural Cues Based on Interaural Coherence." Journal of the Acoustical Society of America 116 (2004): 3075.
- Rose, James, Philip Nelson, Boaz Rafaely, and T. Takeuchi. "Sweet Spot Size of Virtual Acoustic Imaging Systems at Asymmetric Listener Locations." Journal of the Acoustical Society of America 112, no. 5 (2002): 1992–2002.
- Lawrence, M. "Dynamic Range of the Cochlear Transducer." Cold Spring Harbor Symposia on Quantitative Biology (1965).
- Harman, W. M., and B. Rakerd. "Localization of Sound in Rooms IV: The Franssen Effect." Journal of the Acoustical Society of America (1989).
- H. Fletcher and W. Munson. "Loudness, Its Definition, Measurement, and Calculation." The Bell System Technical Journal (1933)
- Robinson, D. W., and R. S. Dodson. "A Re-Determination of the Equal-Loudness Relations for Pure Tones." Journal of Applied Physics 7, no. 5 (1956).
- Olive, Sean. "A Multiple Regression Model for Predicting Loudspeaker Preference Using Objective Measurements: Part 1, Listening Test Results." Presented at the 116th Convention of the Audio Engineering Society, 2004.

- Kleiner, Mendel, and Jiri Tichy. Acoustics of Small Rooms. Boca Raton, FL: CRC Press. ISBN 978-0-203-86924-6.
- Everest, F. Alton, and Ken C. Pohlmann. Master Handbook of Acoustics, 6th ed. New York: McGraw-Hill. ISBN 978-0-07-184103-0.
- Cox, Trevor J., and Peter D'Antonio. Acoustic Absorbers and Diffusers: Theory, Design and Application. London: Spon Press. ISBN 0-415-29649-8.
- Smith, Jim. Get Better Sound. Quarter Note Press. ISBN 0-9820807-0-2.

Made in the USA
Coppell, TX
06 October 2026

92033745R00209